High Acid Crudes

High Acid Crudes

James G. Speight PhD, DSc
CD&W Inc.,
Laramie, Wyoming, USA

AMSTERDAM • BOSTON • HEIDELBERG • LONDON
NEW YORK • OXFORD • PARIS • SAN DIEGO
SAN FRANCISCO • SINGAPORE • SYDNEY • TOKYO
Gulf Professional Publishing is an imprint of Elsevier

Gulf Professional Publishing is an imprint of Elsevier
25 Wyman Street, Waltham, MA 02451, USA
The Boulevard, Langford Lane, Kidlington, Oxford, OX5 1GB, UK

Notices
Knowledge and best practice in this field are constantly changing. As new research and
experience broaden our understanding, changes in research methods or professional practices,
may become necessary.

Practitioners and researchers must always rely on their own experience and knowledge in
evaluating and using any information or methods described herein. In using such information or
methods they should be mindful of their own safety and the safety of others, including parties for
whom they have a professional responsibility.

To the fullest extent of the law, neither the Publisher nor the authors, contributors, or editors,
assume any liability for any injury and/or damage to persons or property as a matter of products
liability, negligence or otherwise, or from any use or operation of any methods, products,
instructions, or ideas contained in the material herein.

Library of Congress Cataloging-in-Publication Data
A catalog record for this book is available from the Library of Congress

British Library Cataloguing-in-Publication Data
A catalogue record for this book is available from the British Library.

ISBN: 978-0-12-800630-6

For information on all Gulf Professional Publishing publications
visit our website at store.elsevier.com

This book has been manufactured using Print On Demand technology. Each copy is produced to
order and is limited to black ink. The online version of this book will show color figures where
appropriate.

Working together
to grow libraries in
developing countries

www.elsevier.com • www.bookaid.org

CONTENTS

Naphthenic acids confer acidic properties on crude oil and the extent of the acidity is expressed as the *total acid number* (TAN), which is the number of milligrams of potassium hydroxide required to neutralize one gram of crude oil. Conversely, the *total base number* (TBN) is determined by titration with acids and is a measure of the amount of basic substances in the oil always under the conditions prescribed by the test method.

High acid crude oils represent the fastest growing segment of global crude oil production. California, Brazil, North Sea, Russia, China, India, and West Africa are known to supply high acid crudes—synthetic crude oil derived from tar sand bitumen is often deemed highly acidic. Due to the overall shortage of processing capacity for high acid crude oil by many refineries and the need to produce low boiling (light) products, high quality crudes are often in high demand, resulting in abundant supply and significant discounts for opportunity crude oils.

The book provides an overview with some degree of detail of the identification of naphthenic acids and their influence on refinery process units. There are process descriptions with a focus on the examination and identification of corrosion and metallurgical problems that occur in process units caused by high acid crude oils.

Laramie, WY

March 15, 2014

CHAPTER 1

Naphthenic Acids in Petroleum

1.1 INTRODUCTION

Crude oil (and the interchangeable term *petroleum*) is a highly complex mixture and typically contains thousands of components (Speight and Ozum, 2002; Hsu and Robinson, 2006; Gary et al., 2007; Cai and Tian, 2011; Speight, 2014a).

Within the large number of constituents of crude oil is a subclass of the oxygen-containing species known as *naphthenic acids* and the term *naphthenic acids* is commonly used to describe an isomeric mixture of carboxylic acids (*predominantly* monocarboxylic acids) containing one or several saturated fused alicyclic rings (Hell and Medinger, 1874; Lochte, 1952; Ney et al., 1943; Tomczyk, et al., 2001; Rodgers et al., 2002; Barrow et al., 2003; Clemente et al., 2003a,b; Zhao et al., 2012). However, in petroleum terminology it has become customary to use this term to describe the whole range of organic acids found in crude oils; species such as phenols and other acidic species are often included in the naphthenic acid category.

Naphthenic acids are a naturally occurring, complex mixture of cycloaliphatic carboxylic acids recovered from petroleum and from petroleum distillates and the term *naphthenic acid*—as used in the petroleum industry—refers collectively to all of the carboxylic acids present in crude oil. Naphthenic acids are classified as monobasic carboxylic acids of the general formula RCOOH, in which R represents the naphthene moiety consisting of cyclopentane and cyclohexane derivatives as well as any acyclic aliphatic acids (Brient et al, 1995; Petkova et al., 2009). Although alicyclic (naphthenic) acids appear to be the more prevalent on the naphthenic acid class, it is now well known that phenol derivatives are also present in crude oil (Speight, 2014).

It has generally been concluded that the carboxylic acids in petroleum with fewer than eight carbon atoms per molecule are almost entirely aliphatic in nature; monocyclic acids begin at

C_6 and predominate above C_{14}. This indicates that the complex structures of the carboxylic acids, which continue to offer challenges in determining these structures, are believed (with reasonable justification) to correspond with those of the hydrocarbons with which they are associated in the crude oil (Robbins, 1998; Rodgers et al., 2002). In the range in which paraffins are the prevailing type of hydrocarbon, the aliphatic acids may be expected to predominate. Similarly, in the ranges in which monocycloparaffins and dicycloparaffins prevail, it has been theorized that the prevalent species will be monocyclic and dicyclic acids, respectively.

More important in the present context, acidic species (naphthenic acids) in the crude oil become active corrosive agents in the distillation column and cause liquid phase corrosion at process temperatures of 250−400°C (480−750°F). Naphthenic acids can cause corrosion in refinery equipment, resulting in costs that are ultimately passed on to the consumer, and the corrosiveness of the acids is believed to be linked to their size and structure. The naphthenic acids content in crude oils is expressed as the total acid number (TAN), which is measured in units of milligrams of potassium hydroxide required to neutralize a gram of oil.

Acidic crude oils are grades of crude oil that contain substantial amounts of naphthenic acids or other acids. They are also called high acid crudes after the most common measure of acidity: the *TAN*. Arbitrarily, a crude oil with a TAN on the order of 0.5 mg KOH/g acid and higher usually qualifies as high acid crude. At an acid number of 1.0 mg KOH/g crude oil, crude oils begin to be heavily discounted in value. Other than acidity, there appear to be no other distinguishing properties that characterize these oils, although most high acid crude oils often have a gravity that is <29°API and are often (but not always) low in sulfur (except for Venezuelan high acid crudes) and frequently produce high yields of gas oil. Acidic oils can vary widely in most other properties.

The interest in (or willingness to accept) high acid crudes as refinery feedstocks high acid crudes is the result of these oils trading at discounts of several dollars per barrel when compared to conventional (low acid) crude oils but processing high acid crude oils (HACs) is also challenging for refineries, especially those not designed to handle crude oil containing naphthenic acids (Heller et al., 1963; Speight, 2014).

In terms of classification, which is not scientifically based, high acid crudes often fall into the subgroup of crude oils known as opportunity crudes. Generally, *opportunity crude oils* are often dirty and need cleaning before refining by removal of undesirable constituents, such as high sulfur, high nitrogen, and high aromatics (such as polynuclear aromatic) components. A controlled visbreaking treatment would *clean up* such crude oils by removing these undesirable constituents (which, if not removed, would cause problems further down the refinery sequence) as coke or sediment.

It is perhaps more correct to separate the *HACs* as a subclass of crude oil. HACs, which contain significant amounts of naphthenic acids, cause corrosion in the refinery—corrosion is predominant at temperatures in excess of 180°C (355°F) (Ghoshal and Sainik, 2013; Speight, 2014)—and occur particularly in the atmospheric distillation unit (the first point of entry of the HAC) and also in the vacuum distillation units. In addition, overhead corrosion is caused by the mineral salts, magnesium, calcium, and sodium chloride which are hydrolyzed to produce volatile hydrochloric acid, causing a highly corrosive condition in the overhead exchangers and must be removed (Erfan, 2011). Therefore, these salts present a significant contamination in opportunity crude oils. Other contaminants in opportunity crude oils which are shown to accelerate the hydrolysis reactions are inorganic clays and organic acids.

1.2 ORIGIN AND OCCURRENCE

Over the next decade (from the time of writing), it is forecast that the supply of high acid crudes (crudes having a TAN in excess of 1.0 mg KOH/g crude oil) will continue to increase significantly, with production rising across the world. All of these crude oils have significant acid numbers. Therefore, corrosion management is of vital importance to ensure that corrosion risk to the plant is minimized and an efficient inspection system must be in place to identify the corrosion which might occur and areas of the plant that might be subject to severe corrosion are identified so that the need for more corrosion resistant alloys can be predicted. In order to understand corrosion and corrosion management, it is necessary to understand the formation (origin) at the time of petroleum generation as well as the nature of naphthenic acids.

The generation of petroleum is associated with the deposition of organic detritus. The detritus deposition occurs during the development of fine-grained sedimentary rocks that occur in marine, near-marine, or even nonmarine. Petroleum is believed to be the product arising from the decay of plant and animal debris that was incorporated into sediments at the time of deposition. However, the details of this transformation and the mechanism by which petroleum is expelled from the source sediment and accumulates in the reservoir rock are still uncertain but progress has been made in environments (Speight, 2014a and references cited therein).

Nevertheless, the composition of petroleum is greatly influenced not only by the nature of the precursors that eventually form petroleum but also by the relative amounts of these precursors (that are dependent upon the local flora and fauna) that occur in the source material. Hence, it is not surprising that petroleum composition can vary with the location and age of the field in addition to any variations that occur with the depth of the individual well. Two adjacent wells are more than likely to produce petroleum with very different characteristics. The same rationale can apply to the occurrence and character of the various constituents of petroleum, not the least of which (in the current context) is the fraction known as *naphthenic acids.*

Naphthenic acid is the generic name used for all of the organic acids present in crude oils—most of the acids arise as biochemical markers of crude oil origin and maturation (Fan, 1991; Speight, 2014). Most of these acids are believed to have the chemical formula $R(CH_2)_nCOOH$, where R is a cyclopentane or cyclohexane ring and n is typically greater than 12 which can lead to the representation of structural characteristics (Tables 1.1–1.3) (Fan, 1991; Hsu et al., 2000; Barrow et al., 2003; Marshall and Rodgers, 2008; Petkova et al., 2009). In addition to $R(CH_2)_nCOOH$, a multitude of other acidic organic constituents are also present in crude oil(s) but not all of the species have been fully analyzed and identified. Although saturated carboxylic acids are the predominant compounds found in most crude oils, aromatic and even heterocyclic compounds have also been reported. For example, monoaromatics and diaromatics have also been identified within the naphthenic acid group (Derungs, 1956; Hsu at el., 2000).

Naphthenic acids are generally described by the formula $C_nH_{2n+Z}O_2$, where n is the number of carbons and Z is the *hydrogen deficiency* index.

Table 1.1 Representative Structures of Naphthenic Acids Found in Crude Oils (Fan, 1991; Hsu et al., 2000; Barrow et al., 2003; Marshall and Rodgers, 2008)

1 ring	2 rings	3 rings	4 rings	5 rings	6 rings	C22 – C33
						$C_nH_{2n-2}O_2$
						$C_nH_{2n-4}O_2$
						$C_nH_{2n-6}O_2$
						$C_nH_{2n-8}O_2$

In each case, R is an alkyl group of varying size and there may be more than one alkyl group per molecule.

Table 1.2 Predominant Naphthenic Acid Compounds Classes in a California Crude Oil (Seifert, 1973)

Table 1.3 Alternate Structural Types of Naphthenic Acids Where the Carboxylic Acid Function Is Attached to an Alkyl Side Chain (US EPA, 2012)

Z is either zero (for simple fatty acids with one carbon–oxygen double bond) or a negative even integer (acids with additional rings/double bonds) that specify homologous series. On a molecular basis, Z is frequently referred to as the molecular *hydrogen deficiency* (Hughey et al., 2002; Qian et al., 2001; Hughey et al., 2007).

Finally, *naphthenates* are the salts of naphthenic acids which have the formula $M(naphthenate)_2$ (M = divalent metal) or are basic oxides with the formula $M_3O(naphthenate)_6$. Naphthenate salts form when naturally occurring naphthenic acids in the crude oil come in contact with metal ions (such as calcium) in the produced water under the right conditions of pH and temperature. Naphthenates ($RCOO-$) are formed during crude oil production due to pressure decrease and release of carbon dioxide, which leads to increase in pH and dissociation of naphthenic acid ($RCOOH$). Naphthenates can precipitate with metal cations present in brine and form deposits, mainly of calcium naphthenates, which can accumulate in topside facilities, desalters, and pipelines, leading to shutdowns and other serious problems during crude refining.

The metal naphthenates are highly soluble in organic media and, with the naphthenic acids themselves, contribute to the interfacial properties of many crude oils and crude oil products (Varadaraj and Brons, 2007; Pillon, 2008).

In addition, under certain conditions, the naphthenic acids present in acidic crude oil will precipitate with calcium ions (Ca^{2+} ions) that are present in produced water and form calcium naphthenate and, to a lesser extent, other metal naphthenates. In fact, calcium naphthenate—a generic term for a deposit which usually contains calcium, sodium, magnesium, iron, and other metal naphthenates, and possibly asphaltene constituents, scale, and other solids—is a troublesome deposit that can form in oil production systems that are handling a crude oil with a high acid number (TAN). The problems caused by calcium naphthenate range from oil treating problems and poor water quality to heavy deposits that can plug lines and valves.

Thermogravimetric analysis has proved to be a suitable tool to investigate calcium naphthenate production and could be used to characterize these solids. The results showed that it is possible to distinguish the compounds produced by specific thermal stability assays. In this report, the synthetic route employed has indicated the formation of different compounds, such as calcium carbonate, calcium sulfate as gypsum, and calcium naphthenate. In the experiments, precipitation of calcium sulfate showed to be dependent on solution pH. Moreover, the reproducibility tests confirmed the qualitative significance in the formation of naphthenates (Moreira and Teixeira, 2009).

1.2.1 Origin

Naphthenic acids are natural constituents of petroleum, where they evolve through the oxidation of naphthenes (cycloalkanes). Initially, the presence of these acidic species was suggested due to process artifacts formed during refining processes, and this may still be the case in some instances. However, it was shown that only a small quantity of acids was produced during these processes (Costantinides and Arich, 1967). Currently, it is generally assumed that acids may have been incorporated into the oil from three different sources: (i) acidic compounds found in source rocks, derived from the original organic matter that created the crude oil (plants and animals), (ii) neo-formed acids during biodegradation (although the high acid concentration in biodegraded oils is believed to be related principally to the removal of nonacidic compounds, leading to a relative increase of the acid concentration levels), and (iii) acids that are derived from the bacteria themselves, e.g., from cell walls that the organisms leave behind when their life cycle is completed (Mackenzie et al., 1981; Behar and Albrecht, 1984; Thorn and Aiken, 1998; Meredith et al., 2000; Tomczyk et al., 2001; Watson et al., 2002; Wilkes et al., 2003; Barth et al., 2004; Kim et al., 2005; Fafet et al., 2008).

This diverse group of saturated monocyclic and polycyclic carboxylic acids can account for as much as 4% (w/w) of crude petroleum (Brient et al., 1995) and represents an important component of the crude oil feedstock as well as waste generated during petroleum processing in some situations (such as the desalting step). Naphthenic acids are also natural constituents of tar sand (oil sand) bitumen and, during the bitumen extraction process, when the alkalinity of the water (pH: approximately 8) promotes solubilization of naphthenic acids (pK_a: approximately 5), the acids are solubilized and concentrated in the tailings stream (Rogers et al., 2001, 2002; Headley and McMartin, 2004; Scott et al., 2005).

In terms of the origin of naphthenic acids, biodegradation of hydrocarbons and the resulting decline in crude oil quality is common in reservoirs cooler than approximately 80°C (176°F). Petroleum biodegrading organisms have a specific order of preference for compounds that they remove from oils and gases (Seifert, 1973; Speight, 2014a). Progressive degradation of crude oil tends to remove saturated hydrocarbons first, concentrating heavy polar and asphaltene components in the residual oil. This leads to decreasing oil quality reflected in a lowering of the API gravity while increasing viscosity, sulfur, and metal contents. In addition to lowering reservoir recovery efficiencies, the economic value of the oil generally decreases with biodegradation owing to a decrease in refinery distillate yields and an increase in vacuum residua yields (Wenger et al., 2001; Speight, 2014). Furthermore, biodegradation leads to the formation of naphthenic acid compounds, which increase the acidity of the oil (typically measured as *TAN*) (Speight and Arjoon, 2012). An increase in the TAN may further reduce the value of the crude oil and may contribute to production and downstream handling problems such as corrosion and the formation of emulsions (see Chapter 3).

Briefly, petroleum biodegradation is the alteration of crude oil caused by living organisms (Conan, 1984). Initially, it was assumed that hydrocarbon degradation only was possible in the presence of oxygen, the processes being carried out by aerobic bacteria (oxygen electron acceptors) (McKenna and Kallio, 1965; Conan, 1984; Waples, 1985). However, anaerobic bacteria also are capable of hydrocarbon degradation in subsurface petroleum reservoirs (Head et al., 2003; Aitken et al., 2004; Huang et al., 2004; Vieth and Wilkes, 2006) and some bacteria can exist under both aerobic and anaerobic conditions (Gaylarde et al., 1999; Grishchenkov et al., 2000; Yemashova et al., 2007). The biodegradation processes are controlled by conditions that support microbial life and suit the specific bacteria, important factors being reservoir temperature, water pH, salinity and nutrient concentrations, and the accessibility to electron acceptors and hydrocarbons (Magot et al., 2000; Peters et al., 2005). Biodegradation has a negative impact on the oil quality and renders both the oil recovery and refining process *difficult − the* molecular changes that take place in the crude oil feedstock as a result of biodegradation the product oil have an increased viscosity over the initial oil.

Naphthenic acids contribute to oil quality debits and may cause additional processing and downstream handling problems. Bacterial activity is strongly controlled by temperature, but it also may be impacted by formation water salinity, availability of free or combined oxygen, and reservoir characteristics.

Evaluating the decline in hydrocarbon quality associated with bio-degradation has become critical in recent years, as offshore drilling has progressed into deeper water depths. In many areas (e.g., offshore West Africa, Brazil, mid-Norway, South Caspian, and eastern Canada), reservoir targets in deep-to-ultra-deep water are shallow, and geothermal gradients are low. These factors make oil quality a major risk because decreased recovery efficiency and oil value compound with higher deep-water operating costs to significantly impact economics, even on major discoveries.

In addition to the concentration of low quality oil components during biodegradation, new compounds can be formed that negatively impact quality. Bacteria appear to manufacture acids, most of which are naphthenic (i.e., cyclic) acids, during the biodegradation of petroleum (Meredith et al., 2000). Because of solubility differences, low molecular weight (C_1-C_5) acids occur predominantly in associated formation waters (Reinsel et al., 1994) while higher molecular weight ($\geq C_{6+}$) acidic species are concentrated in the oil phase. However, apart from this generalization, the distribution of the various naphthenic acid species in crude oil is not fully understood.

Acid contents are usually monitored as a TAN determined by potentiometric titration as per the ASTM D664 method. This method is wrought with potential interferences and interpretive problems (Piehl, 1988) but it still remains a standard method by which oils are assayed and valued. The TAN generally increases with increasing levels of biodegradation. The current activity of biodegrading organisms may be most important in determining organic acid contents because acids may dissipate rapidly owing to relatively high water solubility and reactivity. In addition, this method measures not only the organic acids but also the acidity generated by hydrogen sulfide, carbon dioxide, magnesium chloride ($MgCl_2$), and calcium chloride ($CaCl_2$) in crude oils and may hydrolyze (Piehl, 1960, 1988).

Elevated naphthenic acid contents (TAN >1 mg KOH/g crude oil) are detrimental to crude oil value because acids cause refinery equipment corrosion at high temperatures (Babaian-Kibala et al., 1993, 1998; Turnbull et al., 1998). This can result in an additional valuation debit. Naphthenic acids and their salts (naphthenates) also may lead to the formation of emulsions upon production of biodegraded oils. Sometimes, these emulsions can be tight and difficult to break by conventional means. The additional expense associated with breaking emulsions, especially on production platform sites in deep water, can further challenge field economics. Low molecular weight organic acids in water often impart very foul odors and can cause wastewater disposal problems in refineries processing some biodegraded oils.

1.2.2 Occurrence

HACs have recently become an important development in the international crude oil. The inclusion of substantial amounts of very high acid crudes is projected to take an increasing share of the rising volume of global oil production. Already in excess of 5 million barrels per day of Atlantic basin, crudes have high acid numbers and are sold at a discount to marker crudes that can be substantial (ESMAP, 2005).

High acid crudes are plentiful in most global regions and are increasing their proportion of the total crude supply—South America is a net exporter of high acid crude and high acid crude blends. North West Europe which has traditionally been a net exporter of high acid crudes is now balanced (increased refinery capacity for high acid crudes) and the crudes are exported to the US East Coast, the US Gulf Coast, the Mediterranean area, and even to the Far East. West African high acid crude production is rapidly increasing and being exported out of the region (Table 1.4). Total global supply of high acid crudes is in excess of 9 million barrels per day (Shafizadeh et al., 2010; Gruber et al., 2012; Handa, 2012; Damasceno et al., 2014).

For example, the Central African Republic of Chad ranks as the 10th largest oil reserve holder among African countries, with 1.5 billion barrels of proven reserves as of January 1, 2013 (http://www.eia.gov/countries/country-data.cfm?fips=cd). Crude oil production in

Table 1.4 High Acid Crudes Available to Various Markets	
Source	Crude
North West Europe	Alba
	Captain
	Clair
	Grane
	Gryphon
	Harding
	Heidrun
	Leadon
	Troll blend
South America	Marlim
	Roncador
	Venezuelan blends
West Africa	Ceiba
	Benguela heavy
	Dalia
	Kome
	Kuito
	Lokele
	Rosalita

Table 1.5 QHD Crude Oil Assay (Chevron, 2012a)	
API gravity	16.48
Specific gravity	0.96
Sulfur (% w/w)	0.28
Nitrogen (ppm)	4737.29
Acid number (mg KOH/g)	2.49
Pour point (°C)	− 17.97
Characterization factor (K-factor)	11.76
Viscosity, cSt at 40°C (104°F)	799.76
Viscosity, cSt at 50°C (122°F)	378.71
Vanadium (ppm)	0.51
Nickel (ppm)	13.01
Microcarbon residue (% w/w)	7.11
Ramsbottom carbon (% w/w)	6.11
n-Heptane asphaltene (% w/w)	0.86

Chad was estimated as 115,000 barrels per day (bbl/d) in 2011 and 105,000 bbl/d in 2012, and almost all of this crude oil was exported via the Chad–Cameroon Pipeline. Furthermore, the Chad Doba blend crude oil is substantially affected by this market movement to HAC—Doba blend is a heavy low sulfur, high acid, crude oil (API $\sim 20.6°$, $\sim 0.1\%$ w/w sulfur, TAN ~ 4.7 KOH/mg).

China has seen substantial increases in high acid crude production in recent years and will continue to dominate output through 2015, most likely with substantial increases in production of Qin Huang Dao (QHD) crude oil (Chevron, 2012a) (Table 1.5). In addition, China's National Energy Administration has approved a revised development plan (Conoco Phillips) for second phase development of Penglai 19-3 and Penglai 25-6 oil fields in northern Bohai Bay (Juan and Xian, 2009; Tai and Xian, 2011).

In other regions, Australia, which produced minimal acidic crude output in the Wandoo field, will add substantial high acid crude TAN output with the startup of Vincent and Crosby, both moderately high acid crude grades that will be exported. Duri crude oil (Indonesia) (Table 1.6) (Chevron, 2012b) will increasingly be diverted to domestic use, but remains the closest thing to a regional high acid crude marker (Wu, 2010).

Table 1.6 Duri Crude Oil Assay (Chevron, 2012b)	
API gravity	20.29
Specific gravity	0.93
Sulfur (% w/w)	0.21
Nitrogen (ppm)	3635.70
Acid number (mg KOH/g)	1.46
Pour point (°C)	10.79
Characterization factor (K-factor)	12.13
Viscosity, cSt at 40°C (104°F)	375.74
Viscosity, cSt at 50°C (122°F)	205.40
Vanadium (ppm)	1.35
Nickel (ppm)	39.28
Microcarbon residue (% w/w)	8.01
Ramsbottom carbon residue (% w/w)	7.23
n-Heptane asphaltene (% w/w)	0.08

West African producers have added a number of new HAC since 2004 and overall output of high acid crudes will continue to grow. Angola and Sudan are the main producers of HACs. Sudan will expand substantially the output of high acid crude, particularly for the very acidic Fula crude oil (Dou et al., 2013; Saad et al., 2014). A lack of refining capacity and, in some cases, operational issues, in most of these African market countries will continue to spur export of high acid crudes.

While North Sea overall production will decline, the UK and Norway will experience increases in the output of high acid crudes. Most acidic crude production, such as the Grane crude with an acid number of 2.3 mg KOH/g oil (Statoil, 2012), will continue to be used in NW Europe though the US market receives regular imports of Norwegian cargoes. On the other hand, the output of high acid crudes in Latin America will have an increasing impact on the Asia Pacific crude oil slate—Venezuelan crude oils will continue to appeal to the Chinese and Brazilian for increased sales of high acid crudes. Whether the sales relate to political issues or commercial issues, the next 5 years will see an increase in interregional sales of high acid crudes.

Processing acidic crudes creates corrosion issues to refinery equipment and requires special and expensive technical solutions. Mitigation of process corrosion includes blending, inhibition, material upgrading, and process control (see Chapter 5). Some, but not all, refineries are presently able to refine high acid crudes without suffering serious corrosion problems—to circumvent the effect of high acid crudes, most refiners blend high acid crudes with other crudes before refining. Given that crude is purchased in large parcel sizes, the need to blend one such parcel with multiple parcels of low acid crudes is costly and worthwhile only if there is a substantial discount available for the high acid crude. In the United States, most of the facilities which process heavy sour crude oils and crude oils having a high TAN from Venezuela, Brazil, and Mexico are located on the Gulf Coast. Most of the lighter lower sulfur feedstocks (synthetic crudes) are expected to be shipped to the low- to medium-sulfur refineries in the Mid-Continent, Midwest, and Great Lakes regions of the United States (Baker Hughes, 2010). The refineries which can process high acid crudes are then able to benefit from the existence of the price discount.

The current potential buyers of high acid crudes include: (i) refineries with specialized metallurgy, (ii) large refineries that can dilute acidic crude through blending, (iii) refiners who buy high acid crude when discounted with sufficient specialized facilities to handle these grades, (iv) specialized, non-processing utilizations, (v) risk-adverse refiners who will only occasionally experiment with high TAN grades, and (vi) large refineries buying discounted acidic crude with the plan to use the resid as feedstock for a catalytic cracking unit. Refiners can handle acidic crudes safely through three main methods—dilution (i.e., blending with nonacidic crude), chemical injection, to neutralize acidity, and through the selection and use of specialized materials for the refinery, particularly special alloy steels.

However, the occurrence of naphthenic acids in crude oil is not the only issue when crude pricing is determined. The price differentials between the various crudes depend on a number of factors, including (i) the demand for various petroleum products, (ii) the costs and availability of refineries able to run various crudes to produce the products most in demand, (iii) refinery capacity utilization, (iv) transportation costs, and (v) the relative supply of the different crudes. If the demands for all petroleum products increased proportionately, and all product prices and the general crude price also increased proportionately, then those crudes producing the largest proportion of high value products would increase in price relative to those crudes with a lower proportion of high value products. This effect would be magnified if the demand for the higher value products increased relatively more, so that the price of high value products rose faster than that of low value products.

For example, the diesel fractions derived from Athabasca bitumen and the syncrude oil have low cetane numbers (<30) and very high sulfur content and nitrogen content. Therefore, it is difficult to meet both road cetane and ultralow sulfur specifications on bitumen-derived synthetic crude oil. High nitrogen content also inhibits desulfurization, particularly with respect to dibenzothiophene. In handling these low quality crudes, modifications of the distillate, kerosene, and naphtha hydrotreater must be addressed. The naphtha reformer must also be adjusted since bitumen-derived naphtha has high naphthene content and aromatics content.

1.3 TOTAL ACID NUMBER

Once the presence of acidic species has been determined for crude oil, the next step is to define the crude in terms of acidity. This is done through determination of the acid number, which is measured in units of milligrams of potassium hydroxide required to neutralize a gram of oil. The TAN can be used to further subdefine the crude as *high acid crude* or *low acid crude*.

Regardless of the source, the acids present in the oil cause much corrosion in the refinery equipment. The most common current measures of the corrosive potential of a crude oil are the *neutralization number* or *total acid number*. These are total acidity measurements determined by base titration. Commercial experience reveals that while such tests may be sufficient for providing an indication of whether any given crude may be corrosive, the tests are poor quantitative indicators of the severity of corrosion—for the same TAN, molecular size and structure of the acid also have an important influence (Turnbull et al., 1998).

However, the TAN is expressed in terms of milligrams of potassium hydroxide per gram (mg KOH/g) and is not specific to a particular acid but refers to all possible acidic components in the crude, and is defined by the amount of potassium hydroxide required to neutralize the acids in one gram of oil. A TAN >0.5 mg is (arbitrarily) considered to be high.

As an example, Wilmington and Kern crude oil have a TAN ranging from 2.2 to 3.2 mg KOH/g crude oil, respectively.

However, some acids are relatively inert and, thus, the TAN does not always represent the corrosive properties of the crude oil and, furthermore, different acids will react at different temperatures—making it difficult to pinpoint the processing units within the refinery that will be affected by a particular HAC. Nonetheless, HACs contain naphthenic acids, a broad group of organic acids that are usually composed of carboxylic acid compounds. These acids corrode the distillation unit in the refinery and form sludge and gum which can block pipelines and pumps entering the refinery. The impact of corrosive HACs can be overcome by blending higher and lower acid oils, installing or retrofitting equipment with anticorrosive materials, or by developing low temperature catalytic decarboxylation processes using metal catalysts such as copper. Many California refineries already process high acid crude (Sheridan, 2006).

Although the TAN value is used to define which crudes are high acid crudes, the number is not a reliable indicator of the potential of the crude oil for causing problems with reliability and operations in the refinery. Many acidic species are included in the TAN of the oil and the number actually represents (i) all organic acids, such as naphthenic acids and any other low molecular weight organic acids and (ii) any acids present in the crude that have been added during the production process (Scattergood and Strong, 1987; Craig, 1995, 1996; Hau and Mirabal, 1996; Tebbal and Kane, 1996; Tebbal et al., 1996; Lewis et al., 1999; Kane and Cayard, 2002; Speight, 2014). Typically found are naphthenic acids, which are organic, and also mineral acids such as hydrogen sulfide (H_2S), hydrogen cyanide (HCN), and carbon dioxide (CO_2) can be present, all of which can contribute significantly to corrosion of equipment. Even materials suitable for sour service do not escape damage under such an onslaught of aggressive compounds. Again, because of cost considerations, a trend towards a preference for crude oils with a higher TAN is noticeable.

Current methods for the determination of the acid content of hydrocarbon compositions are well established (ASTM D664) which include potentiometric titration in nonaqueous conditions to clearly defined end points as detected by changes in millivolts readings versus volume of titrant used. A color indicator method (ASTM D974) is also available. However, the ASTM D974 test method is an older method and used for distillates while the ASTM D664 test method is more accurate but measures acid gases and hydrolyzable salts in addition to organic acids. These differences are important on crude oils but less significant on distillates and the naphthenic acid titration (NAT) test method is more precise for quantifying the naphthenic acid content (Haynes, 2006). Both methods (ASTM D664 and ASTM D974) are susceptible to interference for all or any of the following: inorganic acids, esters, phenolic compounds, sulfur compounds, lactones, resins, salts, and additives such as inhibitors and detergents interfere with both methods. Thus, these ASTM methods do not differentiate between naphthenic acids, phenols, carbon dioxide, hydrogen sulfide, mercaptans, and other acidic compounds present in the oil.

- *Potentiometric titration*:
  In this method (ASTM D664), the sample is normally dissolved in toluene and propanol with a little water and titrated with alcoholic potassium hydroxide (if sample is acidic). A glass electrode and reference electrode are immersed in the sample and connected to a voltmeter/potentiometer. The meter reading (in millivolts) is plotted against the volume of titrant. The end point is taken at the distinct inflection of the resulting titration curve corresponding to the basic buffer solution.
- *Color indicating titration*:
  In this test method (ASTM D974), an appropriate pH color indicator (such as phenolphthalein) is used. The titrant is added to the sample by means of a burette and the volume of titrant used to cause a permanent color change in the sample is recorded from which the *TAN* is calculated. It can be difficult to observe color changes in crude oil solutions. It is also possible that the results from the color indicator method may or may not be the same as the potentiometric results.

Test method ASTM D3339 is also similar to ASTM D974, but is designed for use on smaller oil samples. ASTM D974 and D664 use (approximately) a 20 g sample and ASTM D3339 uses a 2.0 g sample. However, both methods use a color change to indicate the end point. ASTM D1534 is designed for electric insulating oils (transformer oils), where the viscosity will not exceed 24 cSt at 40°C (104°F).

The risk with any of these methods is the insufficiency of a specific test method to produce meaningful or realistic data.

For example, the ASTM D974 test method is an older method and used for distillates while the ASTM D664 test method is more accurate but measures acid gases and hydrolyzable salts in addition to organic acids. Moreover, inorganic acids, esters, phenolic compounds, sulfur compounds, lactones, resins, salts, and additives such as inhibitors and detergents may not be amenable to the procedure and may interfere with the method. In addition, many methods do not differentiate between naphthenic acids, phenols, carbon dioxide, hydrogen sulfide, mercaptans, and other acidic compounds present in the oil. In addition, the TAN values as conventionally analyzed and a are no longer considered to be a reliable indicator of corrosivity (Rikka, 2007).

These differences are important for refining crude oil but may be considered to be less significant distillates—the *NAT* method is more precise for quantifying the naphthenic acid content (Haynes, 2006).

The UOP 565 test method (http://www.astm.org/Standards/UOP565.htm) is used to determine the acid number of petroleum products, petroleum distillates, and similar materials by potentiometric titration. Inorganic acids, organic acids, mercaptans, and thiophenols respond to this analysis, but their respective salts do not. For naphthenic acids, it is often preferable to include the salts in the measurement. Therefore, a procedure is also included for the determination of sodium naphthenate salts. The typical range for acid number determination is 0.002–5 mg KOH/g of sample, although higher concentrations can be accommodated.

In addition, there is another test method, the UOP 587 test method (http://www.pngis.net/standards/details.asp?StandardID=UOP + 587% 3A1992) which is used for determining the acid number of petroleum products, petroleum distillates, and other hydrocarbons by colorimetric titration. Inorganic acids, organic acids, mercaptans, and thiophenols respond to this analysis, but their respective salts do not. This method is limited to light colored distillates. For dark colored samples, the potentiometric procedure in UOP Method 565 (http://www.astm.org/Standards/UOP565.htm) is preferred.

The acid number of a sample can be determined on an as-received basis or on a mercaptan and thiophenol-free basis. The method can also be used to determine naphthenic acids, including sodium naphthenates (soaps) in caustic washed hydrocarbons. For an estimated relative molecular mass of 130, the range of detection for naphthenic acids is 5–250 ppm. The latter two procedures apply almost exclusively to low boiling kerosene and low boiling gas oil where it is assumed that the organic acids are entirely naphthenic.

A new method is available for rapid measurement of TAN and boiling point (BP) distribution for petroleum crude and products. The technology is based on negative ion electrospray ionization mass spectrometry (ESI-MS) for selective ionization of petroleum acid and quantification of acid structures and molecular weight distributions. A chip-based nano-electrospray system enables microscale (<200 mg) and higher throughput (20 samples/h) measurement. Naphthenic acid structures were assigned based on nominal masses of a set of predefined acid structures. Stearic acid is used as an internal standard to calibrate ESI-MS response factors for quantification purposes. With the use of structure–property correlations, BP distributions of TANs can be calculated from the composition. The rapid measurement of TAN BP distributions by ESI is demonstrated for a series of high acid crudes and distillation cuts. TANs determined by the technique agree well with those by the titration method. The distributed properties compare favorably with those measured by distillation and measurement of the TANs of corresponding cuts (Qian et al., 2008).

Still another method advocates the use of extraction/esterification/mass spectrometry and provides an analysis of C10 + naphthenic acids. However, lower molecular weight acids can be lost in the esterification step (Jones et al., 2001).

Finally, the *iron powder test* is a method of measuring naphthenic acid corrosion potential and produces results that agree with existing knowledge about this phenomenon (Hau et al., 1999, 2003). The higher the acid content corresponds to greater corrosivity has been observed in the field and in laboratory test results. Advantageously, the iron powder test has revealed that crude oil samples that have the same acid numbers do not show the same corrosivity.

The method has also shown that crude oil samples having higher acid number values can exhibit less corrosivity than other crudes having lower acid numbers—unlike potassium hydroxide, which does not only react with naphthenic acids but also with other compounds such as hydrolyzable salts, iron naphthenates, inhibitors, and detergents, iron powder is more likely to react with all those species also capable of producing corrosion on actual steels. Naphthenic acids present are stronger and expected to produce a larger amount of dissolved iron than in another oil sample having weaker organic acids.

Because of the confusion that can exist when attempting to relate TAN to corrosivity, there has been an interest in developing methods for the analysis of naphthenic acids. For example, the composition of naphthenic acid fraction is helpful in identification of oil source maturation (Headley et al., 2002; Meredith et al., 2000) as well as for fingerprinting fuel spills in the environment (Rostad and Hostettler, 2007). The overall chemical and physical properties of the naphthenic acid fraction (as obtained by extraction from the source) may also vary (Clemente et al., 2003a,b; Headley and McMartin, 2004). Additionally, the need to assess the corrosivity and toxicity including their fate, transport, and degradation has heightened the need for improved characterization of the components of naphthenic acids is necessary since the corrosive and toxic effects are often structure specific (Clemente and Fedorak, 2005).

However, many analytical methods that have been developed to characterize naphthenic acids tend to be semiquantitative, and lack the ability to identify the individual isomers and their respective properties. Thus, a reliable and accurate analytical method is needed to meet the major challenges in devising a suitable test method and these are: (i) quantitation of the total concentration of naphthenic acids in a sample, (ii) characterization of the structures of the compounds in the complex, poorly defined mixtures obtained using various sampling protocols, (iii) determination of the concentration of the true naphthenic acids (as indicated by the definition) and other components found in the mixture obtained following a sampling protocol, and (iv) assessment of the corrosivity (and toxicity) of each of the component types found in the fraction.

1.4 PROPERTIES

The initial observations about naphthenic acids in petroleum and naphthenic acid corrosion date back to the early part of the twentieth century (Ney et al., 1943; Derungs, 1956) and have continued to affect refinery operations to the present time (Speight, 2014b). The so-called *opportunity crudes* with a high naphthenic acid content continue to be exploited (in ever-increasing amounts) and refined (Johnson et al., 2003; Blume and Yeung, 2008). Naphthenic acid corrosion is complicated because of the complexity of the naphthenic acid mixture (and the potential range of structural types in the mixture) (Tables 1.1−1.3) found in crudes from the various sources. Sometimes, the distribution of naphthenic acids is grouped according to their BP, with an implication that naphthenic acids with different BPs lead to different corrosivity (Clemente et al., 2003a,b; Messer et al., 2004).

The prerequisite for designing practical and highly efficient naphthenic acid removal processes (see Chapter 5) is to determine the existing forms, properties, character, and distribution of naphthenic acids (Cai and Tian, 2011). Numerous efforts have been assigned to the determination and analysis of the naphthenic acid fraction of crude oils, with compounds identified including linear fatty acids, isoprenoid acids, as well as monocyclic, polycyclic, and aromatic acids. In addition, other groups of compounds which can influence the acidity of crude oil include inorganic acids, such as some compounds of calcium and magnesium, which are difficult to be removed in the desalting process, and low molecular weight alkyl phenols, which also occur widely in crude oils (Speight, 2014a).

1.4.1 Chemical Properties

Despite the reports of presence of compounds in petroleum acidic fractions that do not solely contain the carboxylic acid functional group, most of the research on petroleum acids comprises the so-called naphthenic acids. The term *naphthenic acids* is commonly used to describe an isomeric mixture of carboxylic acids containing one or several saturated alicyclic rings. Acidic crude oils are generally considered as problematic from an oil quality point of view. The acids cause corrosion problems in the refinery processes and due to their toxicity, they also represent a pollution source in refinery wastewaters.

1.4.1.1 Chemical Structure

Naphthenic acids are a family of carboxylic acid surfactants with varying properties (Tables 1.1−1.3 and 1.7), primarily consisting of cyclic systems that originate from petroleum source material (Brient et al., 1995; Robbins, 1998; Cai and Tian, 2011; Speight, 2014a). The members of this chemical group are composed predominately of alkyl-substituted

Table 1.7 Varying Properties of Naphthenic Acids (Headley and McMartin, 2004)	
Parameter	General Characteristic
Color	Pale yellow, dark amber, yellowish brown, black
Odor	Primarily imparted by the presence of phenol and sulfur impurities; musty hydrocarbon odor
State	Viscous liquid
Molecular weight	Generally between 140 and 450 amu
Solubility	<50 mg/l at pH 7 in water
	Completely soluble in organic solvents
Density	Between 0.97 and 0.99 g/cm^3
Refractive index	Approximately 1.5
pK_a	Between 5 and 6
Log K_{ow} (octanol water partition coefficient)	Approximately 4 at pH 1
	Approximately 2.4 at pH 7
	Approximately 2 at pH 10
Boiling point	Between 250°C and 350°C

Note: All values vary greatly with naphthenic acids source and composition. Values also vary between native and bitumen-extracted compounds.

cycloaliphatic carboxylic acids with smaller amounts of acyclic aliphatic (paraffinic or fatty) acids. Aromatic olefin acids, hydroxyl acids, and dibasic acids are also present as minor components of the naphthenic acid fraction. The cycloaliphatic acids include single rings and fused multiple rings and the carboxyl group is bonded or attached to a side chain or to a cycloaliphatic ring (Dzidic et al., 1988; Fan, 1991).

The components of naphthenic acids are commonly classified by their structures and the number of carbon atoms in the molecule. The polarity and nonvolatility of naphthenic acids increase with molecular weight, which imparts various chemical and physical properties on individual constituents (Robbins, 1998; Headley et al., 2002; Headley and McMartin, 2004; Clemente et al., 2003a,b). However, as a group the naphthenic acids have chemical and physical characteristics that can be used to describe the overall mixture.

1.4.1.2 Acidity

Naphthenic acids confer acidic properties on crude oil and the extent of the acidity is expressed as the *TAN*, which is the number of milligrams of potassium hydroxide required to neutralize one gram of crude oil (ASTM D3339) (Table 1.8). For convenience, HACs, also

Table 1.8 Comments on Testing for Naphthenic Acids	
Origin	Naphthenic acids are natural chemical species occurring in some crude oils
Effects	Naphthenic acids may cause operational problems such as foaming in the desalter or other units
Measurement units	Naphthenic acids are measured as milligrams of KOH per gram of crude oil
Desired levels	The desired level of naphthenic acids in the crude oil is <0.05 mg KOH/g oil
Degree of accuracy	The acid content of a crude oil may be determined to be ± 0.02 mg KOH/g oil, which includes species other than naphthenic acids

called high total acid number (high TAN) crude oils, are crude oils which typically have an overall acidity (expressed as the *TAN*) that exceed some specified arbitrary limit. Typically crudes with a TAN 0.5 mg KOH/g oil or higher are specified as high acid crudes, and they are known to create problems in a refinery. Although not all acidic components in crude are potentially corrosive, refineries find it preferable and less amenable to corrosion if the crude oil feedstock has a TAN <0.5 mg KOH/g crude.

The data for the TAN derived from either (i) potentiometric titration (ASTM D664) or (ii) color indicator titration methods (ASTM D974) are reported in terms of milligrams of KOH per gram of oil sample. The TAN is generally considered to be a measure of all acidic components, including naphthenic acids and sulfur compounds. The TAN has been used to assess the potential for corrosion problems in a crude oil refinery. Values of the TAN in the range of 0.1–3.5 mg KOH/g are common but can be as high as 10.0 for hydrocarbon fractions isolated or produced from crude oil. The basis for the increase in corrosive effects with HAC is presumed to be due to the availability of the carboxylic acid group to form metal complexes.

Conversely, the *total base number* (TBN) is determined by titration with acids and is a measure of the amount of basic substances in the oil always under the conditions of the test (ASTM D2896). Both the TAN and the TBN are a numerical representation of the acidic or basic species in a crude oil. The numbers are not mutually interchangeable in any form and should not be considered to be a reliable guide to the chemical properties and behavior of crude oil.

HACs are usually placed in the class of crude oils known as *opportunity crudes* (Table 1.9), which are typically offered at a discount

Table 1.9 Types of Opportunity Crudes

Crude type	Properties	Concerns
Heavy sour crude	API <26°	Yield slate
	>1 wt% sulfur	Increased delta coke causes lower conversion and higher regeneration temperature in FCC
		Contaminants may harm catalysts
		Corrosion problems in CDU
		Fouling problems
Extra-heavy crude or bitumen	API <10°	Low cetane index for diesel
	Typically also have high metals content	Large amount of resid material
	Viscosity 100–10,000 cP at 60°F (15.6°C) for extra-heavy oil	Large amount of contaminants may harm catalysts
	Viscosity >10,000 cP at 60°F (15.6°C) for bitumen	Fouling problems
High acid crude	TAN >0.5 mg KOH/g	Increased corrosion
	Typically also heavy, API <26°	Poor salt removal and separation of oil and water
		Fouling reduces plant capacity
		Degradation of catalyst activity by calcium
		Low cetane index for diesel
		May impact product specifications

when compared to conventional crude oils and even, on occasion, when compared to heavy crude oil. Opportunity crudes may have various combinations of (i) high sulfur content on the order of >0.7–1.0% (w/w), (ii) high nitrogen content, (iii) high aromatics content, (iv) low API gravity, on the order of <26–28°, (v) high vacuum residuum content, (vi) high viscosity, and (vii) high acidity, with a *TAN* exceeding 0.5 or 1.0 mg KOH/g (Table 1.7).

In some crude oils, the naphthenic acids are the main oxygen-containing components of petroleum (0.5–3.0% w/w), from which they are extracted in the form of salts (naphthenates) by means of an aqueous solution of an alkali. Naphthenic acids are viscous, colorless liquids that turn yellowish upon standing. The BP is on the order of 220–300°C (430–570°F) and the pour point is usually <80°C (176°F). Naphthenic acids are generally insoluble in water but dissolve readily in petroleum products and other organic solvents. Chemically, these constituents exhibit properties that are characteristic of carboxylic acids. Other less desirable properties relate to the corrosivity of these acids.

Furthermore, when high acid crudes are processed, the naphthenic acid corrosion and sulfur corrosion occur together mainly in distilling towers and their adjacent transfer lines. The two corrosive groups, i.e., naphthenic acids and sulfur compounds, influence each other and their effect cannot be simply separated (Laredo et al., 2004). Both are very reactive at high temperatures while naphthenic acid seems to be most aggressive at high velocity encountered in refinery transfer lines. Thus, processing high acid crudes is not always a profitable operation and benefiting from the least expensive crude oils in the market such as high acid crudes, heavy sour crudes, extra-heavy oil, and tar sand (oil sand) bitumen can keep refining margins high. On the other hand, not all refineries can handle crude oil with a high corrosion propensity and a disproportionate amount of bottom-of-the-barrel fractions.

Thus, HACs will have an adverse impact on refinery reliability and operations with corrosion, desalter problems, fouling, catalyst poisoning, product degradation, and environmental discharges. Corrosion caused by such crude oils must be managed by (i) predicting corrosion behavior, (ii) relative risk levels of processing such crudes, (iii) inspection and monitoring methods to identify specific areas that are at risk of naphthenic acid corrosion and to verify that control methods are being used effectively, and (iv) applying suitable corrosion control methods. There are also alloys that appear to be even more corrosion resistant than existing favorites (e.g., 316 SS and 317 SS) (Yeung, 2006). Modifying or removing naphthenic acids will be accomplished via neutralization, decarboxylation, hydrotreating, and extraction (see Chapter 5).

A major concern with processing such crude (and heavy crude oils) is blending and mixtures, since it is common for compatible crudes, such as Louisiana Light-Sweet, West Texas Intermediate, and Alaskan North Slope, from multiple fields to be mixed in pipeline systems, since a lot of refineries are short on crude tankage. The composition of such crude oils is different—incompatibility can occur not only between the easily recognized high asphaltene crude oil and high paraffinic crude oil as well as between HAC and high paraffinic crude oil.

Crude oil incompatibility and the formation of a separate immiscible phase when crude oils are blended can lead to fouling in the desalter, in heat exchangers, and in pipestill furnace tubes. Therefore, it is important to use prerefining test methods (performed by the seller or the buyer) to predict the proportions and order of blending of oils that will prevent incompatibility *prior* to the purchase of the crudes, recognizing that the test method may provide data that include molecular species other than naphthenic acids (Table 1.7). In either case, it is advisable that the buyer be advised by the seller of the quality of the crude oil that is being purchased.

In addition to taking preventative measures for the refinery to process these feedstocks without serious deleterious effects on the equipment, refiners will need to develop programs for detailed and immediate feedstock evaluation so that they can understand the qualities of a crude oil very quickly and it can be valued appropriately and management of the crude processing can be planned meticulously.

In addition, naphthenic acids have been found to cause the formation of soaps. The alkali metals soaps/salts, sodium and potassium naphthenates are water soluble and water dispersible, giving tight emulsions and poor oil-in-water qualities. Naphthenic acid soaps of the alkaline earth metals are insoluble in normal oilfield brines, with a pH >7 at normal upstream process temperatures.

Finally, the toxicity of naphthenic acids is often associated with the surfactant characteristics (Headley and McMartin, 2004). However, since hundreds of the acidic compounds are found in the *naphthenic acid fraction*, it has not conclusively established which specific naphthenic acids are the most toxic. Since no two crude oil reservoirs are precisely exactly the same, the content and complexity of naphthenic acids in crude oils are also not exactly the same. Thus, toxicity does not necessarily correlate directly to the naphthenic acid concentration, but is more a function of content and complexity (Headley and McMartin, 2004).

1.4.2 Physical Properties

Naphthenic acids recovered from refinery streams occur naturally in the crude oil and are not formed during the refining process. Heavy crudes have the highest acid content, and paraffinic crudes usually have low acid content. Although the presence of naphthenic acids has been established in almost all types of crude oil, only certain naphthenic- and asphalt-based crudes contain amounts that are high enough to require treatment in order to meet product specifications (US EPA, 2012).

Naphthenic acids are obtained by caustic extraction of petroleum distillates, primarily kerosene and diesel fractions. In addition to reducing corrosion in the refinery, the caustic wash of the distillates is necessary to improve the technical properties, storage stability, and odor of the finished kerosene and diesel fuels. The commercial production of naphthenic acid from petroleum is based on the formation of sodium naphthenates which occurs when the petroleum distillates are treated with sodium hydroxide caustic. Since this reaction occurs *in situ*, *sodium salts of naphthenic acids are* considered an intermediate stream in the production of refined naphthenic acid. The sodium naphthenate-containing solutions contain approximately 5−15% (w/w) sodium naphthenate, 0−0.5% (w/w) sodium mercaptide (RS^-Na^+), and 3−4% (w/w) sodium hydroxide in water with a highly alkaline pH (pH >12). These caustic solutions are typically sent to specialized facilities in which they undergo further processing to recover the naphthenic acids.

1.4.2.1 Melting Point
Because naphthenic acids are not pure chemicals, the melting point characteristics of these complex substances vary with the hydrocarbon composition of their make-up. Based on data available in commercial product specifications and Material Safety Data Sheets (MSDS), substances produced for commercial use have melting points that fall in the range from −35°C to +2°C (−31 to 36°F) (US EPA, 2012).

1.4.2.2 Boiling Point
Because these substances are not pure chemicals, the BP characteristics of naphthenic acids and their salts vary according to the hydrocarbon component make-up of the complex substances in which they are found. Based on data available in commercial product specifications and MSDS, substances produced for commercial use have BPs that fall in the range from 140 to 370°C (285 to 700°F) (US EPA, 2012).

1.4.2.3 Solubility
Naphthenic acids can be very water soluble to oil soluble depending on their molecular weight, process temperatures, salinity of waters, and fluid pressures. In the water phase, naphthenic acids can cause stable reverse emulsions (oil droplets in a continuous water phase). In the oil phase with residual water, these acids have the potential to react with a host of minerals, which are capable of neutralizing the acids. The main reaction product found in practice is the calcium naphthenate soap (the calcium salt of naphthenic acids).

Chemically, naphthenic acids are weak acids having pK_a values of approximately 5–6 (Brient et al., 1995; Havre, 2002). As the pH of a solution of naphthenic acids increases above the pK_a value, a greater proportion of the constituents are ionized and exist in the dissolved phase of the aqueous medium (Havre, 2002). Therefore, alkaline solutions increase a naphthenic acid's solubility, and acid solutions decrease solubility (Havre, 2002). Product literature references have cited narrative statements such as very low water solubility or only slightly soluble in water.

1.4.2.4 Interfacial Properties

Crude oil components that are surface active (at the boundary between liquid/solid or liquid/gas) and interfacial active (in the interface between two liquids) span over a large range of chemical structures and molecular weights. Asphaltene constituents and resin constituents, including naphthenic acids, are examples of petroleum constituents that display such properties, thus acting as natural surfactants (Seifert and Howells, 1969; Hoeiland et al., 2001; Langevin et al., 2004; Poteau et al., 2005). Surfactant molecules are amphiphilic, meaning that they have both hydrophilic and hydrophobic parts in the molecules, and for this reason adsorb strongly at interfaces (Pashley and Karaman, 2004). Oil/water interfacial active compounds typically reduce the interfacial tension between the two phases and enhance stabilization of water-in-oil (w/o) emulsions. The formation of stable w/o emulsions is generally undesirable and causes serious challenges in petroleum production in terms of separation and refining processes (Sjöblom et al., 2003; Bennett et al., 2004; Ese and Kilpatrick, 2004).

Surface active naphthenic acids in crude oils are also important for reservoir production challenges (Sjöblom et al., 2003; Bennett et al., 2004; Ese and Kilpatrick, 2004). Although the processes involved are complex and still not well understood, polar compounds present in crude oils are generally assumed to be involved in adsorption interactions or deposition mechanisms that take place at the crude oil/brine/rock interface. These surface active components may contribute to wettability alteration of the rock from initial water wet to less water wet or oil wet reservoirs.

1.4.2.5 Environmental Effects

Naphthenic acids are toxic to aquatic algae and other microorganisms—naphthenic acid molecules possess hydrophilic and hydrophobic functional groups which allow these molecules to penetrate into cell membranes and disrupt cellular function, eventually resulting in cell death (MacKinnon and Boerger, 1986; Frank et al., 2008, 2009) and it has been shown (Herman et al., 1994) show that acute toxicity of tar sand (oil sand) process water by natural processes is reduced within 1 year while the removal of chronic toxicity requires 2 to 3 years.

The degradation and detoxification rates have been shown to be related to structure and the same relationship might be expected for corrosion and structure. Thus, toxic effects do not relate directly to the concentration of naphthenic acids but are more a function of content and complexity of naphthenic acid constituents (Brient et al., 1995; Lai et al., 1996; Rogers et al., 2002). Unfortunately, because of the inability of the various test methods and test protocols to differentiate between the individual structures of the naphthenic acid constituents, it cannot be conclusively established which specific naphthenic acids are the most toxic (and the most corrosive) due mainly to the presence of hundreds of these compounds in the naphthenic acid fraction.

1.4.2.6 Biodegradation
Naphthenic acids are amenable to microbial utilization similar to other hydrocarbon compounds—certain microorganisms are capable of degrading complex mixtures of commercial sodium salts of naphthenic acids as well as mixtures of extracted naphthenic acids (Herman et al., 1993, 1994; Clemente et al., 2004; Clemente and Fedorak, 2005).

Although rates of biodegradation may be affected by steric factors related to the numbers of cycloalkane rings or the alkyl constituents on the ring structure, microbial populations respond to naphthenic acid substrates through increased carbon dioxide production, oxygen consumption, and enhancement of metabolism with the addition of nutrients. With single ring naphthenic acids, biodegradation of both the ring and side chain acid has been shown to occur (Herman et al., 1993, 1994). As the number of cycloalkane rings increase, it may be inferred from what is known about degradation of multi-ring naphthenes that biodegradation rates may slow, but these substances will degrade given time (Bartha and Atlas, 1977; Clemente et al., 2004; Clemente and Fedorak, 2005).

CHAPTER 2

Mechanism of Acid Corrosion

2.1 INTRODUCTION

Corrosion is the deterioration or destruction of metals and alloys in the presence of an environment by chemical or electrochemical means (Chapter 1).

Corrosion is an irreversible interfacial reaction of a material (metal or ceramic or polymer) with its environment which results in its consumption or dissolution into the material of a component of the environment. Often, but not necessarily, corrosion results in effects detrimental to the usage of that material considered. Exclusively physical or mechanical processes such as melting and evaporation, abrasion or mechanical fracture are not included in the term corrosion.

Corrosion is a natural phenomenon and is the deterioration of a material as a result of its interaction of the material with the surroundings (Fontana, 1986; Garverick, 1994; Shreir et al., 1994; Jones, 1996; Shalaby et al., 1996; Peabody, 2001; Bushman, 2002; Landolt, 2007). Although this definition is applicable to any type of material, it is typically reserved for metallic alloys (Speight, 2014b). Furthermore, corrosion processes not only influence the chemical properties of a metal or metals alloy but also generate changes in the physical properties and the mechanical behavior.

Corrosion is an ever-present phenomenon in refineries but the extent of the corrosion is dependent upon the characteristics of the refinery feedstocks and the nature of the processes and the process parameters. When uncorroded steel is exposed to the air (an oxidizing atmosphere), the originally unadulterated (and somewhat shiny) surface of the steel will eventually (depending upon the ambient conditions) be covered with rust (an oxidation product of the iron component of the steel) and the steel is stated to be *corroded*. Chemically, the tendency of metals to corrode is related to the low stability of the metallic (zerovalent or zero oxidation) state. When metals occur in the form of compounds with other elements (they acquire positive states of oxidation)—such is the case with the formation of iron oxide (rust).

Generally, corrosion of iron in an oxidizing atmosphere is represented as the formation of rust (iron oxide) as might occur on the exterior surfaces of reactors and pipelines (Beavers and Thompson, 2006) which is often represented by the formation of iron oxide and then hydration in humid environments:

$$2Fe^0 + 3O_2 \rightarrow 2Fe_2O_3$$
$$2Fe_2O_3 + xH_2O \rightarrow 2Fe_2O_3.xH_2O$$

In reality, rust formation is a much more complex process and can occur at the point of, or at some distance away from, the actual pitting or erosion of iron (Speight, 2014b). The involvement of water accounts for the fact that rust formation occurs much more rapidly in moist conditions than in a dry environment (such as a desert) where water in the atmosphere or water in the ground is limited.

However, acid corrosion does not follow the same type of chemistry and is the deterioration a material undergoes as a result of interactions of the material with the surrounding acidic environment and, in the current context of naphthenic acids, is interior corrosion of the reactor or pipeline (Fontana, 1986; Shreir et al., 1994; Jones, 1996; Peabody, 2001; Bushman, 2002; Landolt, 2007). Although this definition is applicable to any type of material, it is usually reserved for metallic alloys of the types found in industrial settings—such as the petroleum industry. Approximately 80 of the known chemical elements are metals (Speight, 2014b), and approximately half of the metals can be alloyed with other metals, giving rise to several thousand different alloys—many of which are used in refineries. Furthermore, each of the alloys will have different physical, chemical, and mechanical properties, but all of the alloys can corrode to some extent and in different ways due to attack by acidic species. Furthermore, corrosion of an alloy (or metal) influences the chemical properties of the alloy (or metal) and also causes changes in the physical properties and mechanical properties of the alloy which influence behavior and longevity of the alloy as they relate to in-use performance.

2.2 TYPES OF CORROSION

Generally, corrosion can be classified into three general forms based on the type of damage that results and the three general forms are: (1) uniform corrosion, (2) localized corrosion, and (3) stress corrosion cracking. Some types of corrosion damage can be tolerated while other corrosion damage cannot be tolerated and it is important to be aware of these distinctions.

Irrespective of the type of corrosion that occurs, naphthenic acids in crude oil can cause corrosion which often occurs in the same places as high-temperature sulfur attack such as heater tube outlets, transfer lines, column flash zones, and pumps. Furthermore, naphthenic acids alone or in combination with other organic acids (such as phenols) can cause corrosion at temperatures as low as 65°C (150°F) up to 420°C (790°F) (Gorbaty et al., 2001; Kittrell, 2006). Crude oil with a total acid number (TAN) higher than 0.5 mg KOH per gram of crude oil and crude oil fractions with a TAN higher than 0.5 mg KOH per gram of crude oil fractions are considered to be potentially corrosive between the temperature of 230−400°C (450−750°F).

Some other forms of corrosion are galvanic corrosion, selective alloy breakdown, intergranular corrosion, fatigue, friction, erosion, cavitation, hydrogen embrittlement, biocorrosion, and high-temperature oxidation. Since these forms of corrosion have been described in detail elsewhere (Speight, 2014b), only the three general forms of corrosion enumerate above (i.e., uniform corrosion, localized corrosion, and stress corrosion cracking) will be dealt with here.

2.2.1 Uniform Corrosion

The most common form of corrosion is *uniform corrosion*, whereby there is a generalized, overall chemical attack of the entire exposed surface of the metal, leading to a more or less uniform reduction in the thickness of the affected metal. *Uniform corrosion*, in which metal is removed more or less uniformly, is the most common form of corrosion and the least dangerous. It is generally agreed that the maximum acceptable loss of metal due to uniform corrosion is approximately 20 mils per year (mpy). This rate of corrosion is not usually desirable since high corrosion rates not only reduce the thickness of piping but also can lead to plugging of heat exchanger bundles and reactor screens by corrosion deposits. For example, sulfidic corrosion, which can occur when naphthenic acids and sulfur are present in the feedstock, results in the formation of iron sulfide (Bota and Nesic, 2013) and the iron sulfide scale occupies a volume approximately seven times the volume of metal that is removed, thus a 10 in. (internal diameter) pipe corroding at 20 mpy would produce approximately 3 ft^3 of loose scale per year per 100 ft of pipe length. In the presence of acidic corrosion, this phenomenon would cause pipe failure in a very short time.

2.2.2 Localized Corrosion

In contrast to uniform corrosion, there is the process of *localized corrosion* in which an intense attack takes place only in and around particular zones of the metal, leaving the rest of the metal unaffected—an example is pitting corrosion. *Localized corrosion* involves selective removal of metal from part of the exposed metal surface. Pitting corrosion, crevice corrosion, galvanic corrosion, and selective weld attack all fall under this category. These types of damage are difficult to inspect and, unlike uniform attack, increased corrosion allowances are seldom an effective control measure.

Corrosion by naphthenic acids typically has a localized pattern (*localized corrosion*), particularly at areas of high velocity and, in some cases, where condensation of concentrated acid vapors can occur in crude oil distillation units (Hilton and Scattergood, 2010). The attack also is described as lacking corrosion products—as opposed to oxidative corrosion (which can form rust) or sulfidic corrosion (which can result in the formation of iron sulfide flakes). Damage is in the form of unexpected high corrosion rates on alloys that would normally be expected to resist sulfidic corrosion (particularly steels with more than 9% chromium). In some cases, even very highly alloyed materials (i.e., 12% chromium, type 316 stainless steel (SS) and type 317 SS), and in severe cases even 6% Mo (molybdenum) stainless steel has been found to exhibit sensitivity to corrosion under conditions as often exist in the distillation unit.

2.2.3 Stress Corrosion Cracking

On the other hand, *stress corrosion cracking* is another form of corrosion caused by the presence of naphthenic acids in the feedstock. This form of corrosion involves cracking of metal without significant loss of metal and occurs when certain metals are exposed under a tensile stress to specific environments and failures can occur rapidly without warning, thus it is important that the risk be minimized. Stress corrosion cracking can be prevented by: (1) selecting metals which are immune to failure when attached by naphthenic acids—this is usually the preferred method, (2) removal or reduction of stress, or (3) control of the environment.

In the case of naphthenic acid corrosion, removal of the acidic species from the feedstock (Chapter 5) can resolve items (2) and (3).

2.3 CORROSION BY ACIDIC SPECIES

Acid corrosion is typically (but not always) due to the presence of acidic constituents in crude oil. These are the so-called naphthenic acids and various other acidic species that fall within the acid group and which arise as biochemical markers of crude oil origin and various maturation processes (Chapter 1). Naphthenic acid corrosion, which can also include corrosion by other acidic species in the oil, represents an important challenge for the oil refining industry when high acid and/or opportunity crudes (low-quality crude oils) are processed. The corrosivity of low-quality crudes is caused mainly by their natural naphthenic acids (NAP) and sulfur content, which becomes particularly problematic at high-temperature and high-velocity conditions, typical for distilling towers, furnaces, and transfer lines, causing important material losses during processing.

Naphthenic acid is the generic name used for all of the organic acids present in crude oils (Chapter 1)—most of the acids arise as biochemical markers of crude oil origin and maturation (Fan, 1991). Most of these acids are believed to have the chemical formula $R(CH_2)_nCOOH$, where R is a cyclopentane ring or a cyclohexane ring and n is typically greater than 12. In addition to $R(CH_2)_nCOOH$, a multitude of other acidic organic constituents are also present in crude oil(s) but, in spite of many claims to the contrary, not all of the species have been fully analyzed and identified and may never by fully identified (*never* being a long time but is used here to be illustrative of the monumental task required) (Chapter 1) (Blanco and Hopkinson, 1983; Fan, 1991; Babaian-Kibala et al., 1999; Tebbal and Kane, 1996; Tebbal et al., 1997; Tebbal and Kane, 1998; Tebbal, 1999; Dettman et al., 2010).

Naphthenic acid corrosion is one of the well known and serious problems in the petroleum refining industry (Derungs, 1956; Gutzeit, 1977; Slavcheva et al., 1998, 1999; Qu et al., 2005, 2006). The rates of naphthenic acid corrosion do not always increase with an increase in the TAN but do increase with increasing temperature (Chapter 3). In order to obtain the credible corrosion rate, comprehensive analysis and estimation should be performed on the feedstock on the basis of considering not only the temperature but also any other relevant factors (Dettman et al., 2010; Ayello et al., 2011; Wang et al., 2011a). In the petroleum refining industry and the gas processing industry, factors associated with the composition and behavior of the feedstocks (such as temperature, TAN, fluid velocity, gas velocity, and pipeline/reactor material) work simultaneously and cumulatively.

At oil processing temperatures, naphthenic acids show corrosion activity and although there has been numerous works to determine the specific factors, the nature of naphthenic acid corrosion and the factors controlling it are still not completely understood (Kane and Cayard, 1999, 2002; Wu et al., 2004a, 2004b; Flego et al., 2013). There are two major reasons for this general lack of understanding (Kane and Cayard, 1999, 2002; Wu et al., 2004a, 2004b): (1) the extreme complexity and interplay of the factors affecting the corrosion and erosion—corrosion processes such as the different corrosivity of crude oils, depending on the TAN of the crudes as well as the naphthenic acid activity and their distribution over boiling points and decomposition temperatures plus the presence of additional corrosion-active compounds, such as organic sulfur compounds and chlorides, in crude; (2) the variable refining process parameters, such as the hydrocarbon feedstock flow rate, the extent of oil evaporation, and the processing temperature; and (3) the susceptibility of metal equipment to corrosion. The second reason is the lack of laboratory units for effective restored-state experiments mimicking the actual high-temperature and high-feedstock flow rate conditions.

Examples of the variation in the properties of crude oils rich in naphthenic acid species are: Captain, Alba, Gryphon, Harding and Heidrun crudes are all considered opportunity crudes from the North Sea region. The specific gravity of these crudes ranges from 0.88 to 0.94 and the TANs range from 2.0 to 4.1. Gryphon has a sulfur content of 0.4% w/w while Alba has 1.3% w/w sulfur. Such variations require refiners to find high-acid crude oils (HACs) that are suitable for their processing and product profiles—not always an easy task and accurate analysis for estimation of crude oil behavior in the refinery is essential.

Naphthenic acid corrosion is differentiated from sulfidic corrosion by the nature of the corrosion (pitting and impingement) and by the severe (naphthenic acid) corrosion at high velocity in distillation units.

Feedstock heaters, furnaces, transfer lines, feed and reflux sections of columns, atmospheric and vacuum columns, heat exchangers, and condensers are among the types of equipment subject to this type of corrosion. Furthermore, isolated deep pits in partially passivated areas and/ or impingement attack in essentially passivation-free areas are typical of naphthenic acid corrosion. In many cases, even very highly stable alloys (i.e., 12 Cr, AISI types 316 and 317; AISI is American Iron and Steel Institute) have been found to exhibit sensitivity to corrosion under these conditions.

The acidic species in crude oil can be semiquantified (even pseudo-quantified) by the TAN of the crude oil (Chapter 1) (Scattergood and Strong, 1987; Craig, 1995, 1996; Hau and Mirabal, 1996; Tebbal and Kane, 1996; Lewis et al., 1999; Kane and Cayard, 2002), which is expressed in terms of milligrams of potassium hydroxide per gram (mg KOH/g oil). However, the TAN is not specific to particular acid constituents but refers to all possible acidic components of the oil and is the amount of potassium hydroxide required to neutralize the acids in 1 g of oil; alternatively, the base number can also be used (ASTM D974, ASTM D1386, ASTM D2896, ASTM D3242, ASTM D3339, ASTM D4739, ASTM D5770, ASTM D7253, ASTM D7389).

Although the TAN (Chapter 1) was used to refer to organic naphthenic acids that belonged to the acid group (-CO_2H), other organic acids (such as derivatives of phenol) and mineral acids such as hydrogen sulfide (H_2S), hydrogen cyanide (HCN), and carbon dioxide (CO_2) also contribute significantly to the TAN and hence to the corrosion of equipment. HACs, also called high-total acid number (high-TAN) crude oils, are oils which typically have a TAN in excess of 0.5—this assignment based on the total number acid is arbitrary since corrosivity is crude oil specific and process specific. Nevertheless, processing HACs is also challenging for refineries, especially those not designed to handle crude oil containing naphthenic acids (Heller et al., 1963).

Besides sulfur, crude contains many species that are quantified by the TAN of the oil (Scattergood and Strong, 1987; Craig, 1995, 1996; Hau and Mirabal, 1996; Tebbal and Kane, 1996; Lewis et al., 1999; Kane and Cayard, 2002; Speight, 2009). The TAN number is expressed in terms of milligrams of potassium hydroxide per gram (mg KOH/g) and is not specific to a particular acid but refers to all possible acidic components in the crude and is defined by the amount of potassium hydroxide required to neutralize the acids in 1 g of oil. Typically found are naphthenic acids, which are organic, but also mineral acids such as hydrogen sulfide (H_2S), hydrogen cyanide (HCN), and carbon dioxide (CO_2) can be present, all of which can contribute significantly to corrosion of equipment. Even materials suitable for sour service do not escape damage under such an onslaught of aggressive compounds. Again, because of cost considerations, a trend toward a preference for crude oils with a higher TAN is noticeable.

HACs, also called high-TAN crude oils, are oils which typically have a TAN number in excess of 0.5. HAC trade at discounts of about $3/bbl to $10/bbl to conventional (low acid) crude oils but processing HACs is also challenging for refineries, especially those not designed to handle crude oil containing naphthenic acids (Heller et al., 1963).

Naphthenic acid corrosion is one of the serious long-known problems in the petroleum refining industry (Derungs, 1956; Gutzeit, 1977; Slavcheva et al., 1998, 1999; Qu et al., 2005, 2006). The rates of naphthenic acid corrosion increase with the temperature rising and are also influenced to some extent by the acid content as determined by the TAN. Under the condition of low velocity of naphthenic acid, the influences of fluid velocity on corrosion rate can be detected, especially in the high-temperature conditions. In petroleum refining industry, influence factors (such as temperature, TAN, fluid velocity, and material) work simultaneously. In fact, the TAN is not useful as a characteristic for determining of corrosivity anymore. In this context, methods for the exact analysis of crude oil naphthenic acids are becoming of great importance, as well as revealing the correlation between the configuration and functionality of acids and their corrosivity. Thus, in order to obtain the credible corrosion rate, the comprehensive analysis and estimation should be done on the basis of considering the combining temperature with other factors and not only by the TAN but also the structure of the naphthenic acid constituents as well as sulfur content of the crude oil and other factors related to the reactor parameters (Slavcheva et al., 1998, 1999; Meredith et al., 2000; Laredo et al., 2004).

High-temperature corrosivity of crude oil as illustrated by corrosion in distillation units is a major concern of the refining industry. The presence of naphthenic acid and sulfur compounds considerably increases corrosion in the high-temperature parts of the distillation units and, therefore, equipment failures have become a critical safety and reliability issue.

Isolated deep pits in partially passivated areas and/or impingement attack in essentially passivation-free areas are typical of naphthenic acid corrosion. Damage is in the form of unexpected high corrosion rates on alloys that would normally be expected to resist sulfidic corrosion. In many cases, even very highly alloyed materials (i.e., 12 Cr, AISI types 316 and 317) have been found to exhibit sensitivity to corrosion under these conditions. Naphthenic acid corrosion is differentiated from sulfidic corrosion by the nature of the corrosion (pitting and impingement) and by its severe attack at high velocities in crude distillation units. Crude feedstock heaters, furnaces, transfer lines, feed and reflux sections of columns, atmospheric and vacuum columns, heat exchangers, and condensers are among the types of equipment subject to this type of corrosion.

Crude corrosivity has been loosely associated as a function of the TAN but, among other (unknown) factors, the sulfur content, with particular emphasis on the sulfur type present in the crude oil, also plays a role in crude oil corrosivity. Generally, crude oils that have high acid numbers and low sulfur content are particularly corrosive. In fact, it is possible to develop a series of operating envelopes in terms of TAN and sulfur content (particularly, the corrosive sulfur compounds). For each operating envelope, a specific corrosion control action is required. The economically best option is to run to the limit of a chosen corrosion control level. This approach maximizes the amount of corrosive crude that can be processed for a given level of corrosion control.

Experience has shown that naphthenic crudes generally first affect the vacuum transfer line and VGO side-stream. As feed TAN is further increased, corrosion will affect the atmospheric transfer line and heavy atmospheric gas oil (HAGO) circuits and the bottoms of both towers. Furthermore, while high-acid crudes primarily affect the hot parts of the atmospheric distillation unit and the vacuum distillation unit, they also affect downstream units as well (e.g., hydrotreater preheat trains). Some crude oils have also caused overhead system corrosion because of poor (inefficient) salt removal in the dewatering/desalting unit prior to introduction of the feedstock to the distillation units.

Concurrent presence of both sulfur-containing compounds and naphthenic acids in crudes introduces a new level of complexity into the study of the corrosion mechanisms. Sulfidic corrosion leads to formation of a sulfide scale, which provides some degree of protection against naphthenic acid corrosion.

2.3.1 Chemistry

Corrosion processes caused by the presence of naphthenic acids in the feedstock result in the formation of iron naphthenates in which the acid moieties react with the iron in the metal (or alloy) from which the reactor is constructed:

$$Fe + 2RCOOH \rightarrow Fe(RCOO)_2 + H_2 \tag{1}$$

$$Fe(RCOO)_2 + H_2S \rightarrow FeS + 2RCOOH \tag{2}$$

$$Fe + H_2S \rightarrow FeS(oil\ insoluble) + H_2 \tag{3}$$

The first reaction (Reaction 1) produces the oil-soluble iron (iron naphthenate) and the second reaction inhibits soluble iron production—each reaction is dependent upon the functionality of the sulfur compounds (Meriem-Benziane and Zahloul, 2013). If sulfur is present (as is often the case), the reaction proceeds even further with the formation of iron sulfide (Reaction 2). In short, the amount of sulfur in a crude oil (determined as total sulfur, S% w/w) is not necessarily related to reactivity—for example, hydrogen sulfide is very reactive toward iron, producing a protective layer of iron sulfide (FeS) that can prevent further corrosion by acidic species.

The first reaction (Reaction 1) is the direct attack of naphthenic acid species on the steel lining (low-carbon steel) of columns or reactors and the second reaction (Reaction 2) is responsible for hydrogen sulfide corrosion. Iron naphthenate as a corrosion product is well soluble in oils, and iron sulfide creates a protecting film on the metal surface (Reaction 2). In some instances, this film may not form unless the sulfur content in oil is at least 2—3% w/w (Jayaraman et al., 1986). By Reaction (3), dissolved iron naphthenate interacts with hydrogen sulfide to regenerate naphthenic acids and to produce iron sulfide, which precipitates in the oil thereby affecting its quality. Yépez (2005) considered the influence of sulfur compounds on the processes occurring during naphthenic acid corrosion. It was shown that the inherent presence of these compounds in oils can have directly opposed consequences of NAC: (1) when the system contains sulfur compounds that produce hydrogen sulfide, a protective FeS film is formed on the metal surface (and the further corrosion of metal is stopped); (2) if the system contains sulfoxides, water forms in the cathode zone and the corrosion processes are intensified. The results of the cited study emphasize that it is important to control the presence of sulfoxides in crude oils because these compounds pose a hazard as promoters of petroleum acid corrosion.

Since the iron naphthenates are soluble in crude oil, the surface of the metal is relatively film free. Furthermore, alkali-naphthenates accelerated the naphthenic acid corrosion, and with an increase in the alkali metal atom radius, the corrosion acceleration enhanced. Sodium naphthenates can remove the corrosion product from the metal surface, thus accelerating the corrosion process (Wang et al., 2011a, 2011b).

However, the presence of both naphthenic acids and sulfur-containing compounds in crude oil introduces a new level of complexity into the chemistry of corrosion. Sulfidation corrosion of the metal (typically iron) leads to formation of a metal sulfide (typically iron sulfide) scale, which provides some degree of protection against the onset of naphthenic acid corrosion (Kanukuntla et al., 2008).

In the presence of hydrogen sulfide, a sulfide film is formed which can offer some protection depending on the acid concentration. If the sulfur-containing compounds are reduced to hydrogen sulfide (H_2S), the formation of a potentially protective layer of iron sulfide occurs on the reactor walls and corrosion is reduced (Kane and Cayard, 2002; Yépez, 2005; Bota and Nesic, 2013). When the reduction product is water instead of or as well as hydrogen sulfide, coming from the reduction of sulfoxide derivatives, the naphthenic acid corrosion is enhanced (*wet corrosion*) (Yépez, 2005).

Furthermore, the accumulation of corrosive hydrogen sulfide in a reactor enhances corrosion as opposed to diminished corrosion in a flow-through reactor. This effect (corrosion by hydrogen sulfide) is aggravated with time and particularly for crude oil that has higher total sulfur content and at higher temperature. In addition, and contrary to early theories, corrosion rates (which are, because of other factors, feedstock dependent) tend to remain constant with increasing TAN until a critical value of the TAN is reached—the so-called critical value is a function of the oil total sulfur content. At higher TANs, the corrosion rate increases sharply marking a transition from sulfidation to naphthenic acid dominated corrosion regime (Kanukuntla et al., 2008).

During high-temperature processes—such as the distillation process during which the temperature of the crude oil in the distillation column can be as high as 400°C—thermal decarboxylation of naphthenic acids can occur:

$$R\text{-}CO_2H \rightarrow R\text{-}H + CO_2$$

Carbon dioxide, which is not altogether a noncorrosive material, may show some signs of corrosivity (depending on the conditions when it is formed), and the presence of water in the system results in the formation of carbonic acid, which has enhanced corrosivity over carbon dioxide:

$$CO_2 + H_2O \rightarrow H_2CO_3$$

In fact, organic acids, such as acetic acid (CH_3CO_2H), enhance the corrosion rate of mild steel by accelerating a cathodic (electrochemical) reaction (Garsany et al., 2002; Dougherty, 2004; Matos et al., 2010; Amri et al., 2011; Tran et al., 2013; Speight, 2014b). This is the same type of corrosion enhancement that might be expected when naphthenic acids are present, especially when the degradation products of the naphthenic acids are lower molecular weight fatty acids (such as formic acid and acetic acid), which are also corrosive (Gutzeit, 1977). However, the precise chemical aspects of the reaction remain unresolved (some would say *controversial*), and it is not clear whether the adsorbed acetic acid molecule is reduced at the surface (in addition to any reduction of hydrogen ions) (*direct reduction*).

The alternative possibility is that the acetic acid dissociates and provides an additional source of hydrogen ions (H^+) near the steel surface, while the only cathodic reaction is reduction of hydrogen ions, and induces a chemical mechanism referred to as a *buffering effect*—an effect in which one chemical prevents chemical changes in a chemical system. In some chemical reactions, the buffers are added to the mix or are formed naturally as a result of, or during, the reaction progress.

Such an effect often adds chemical confusion to attempts to assign precise chemistry to the mechanism of corrosion by naphthenic acids, leaving the reaction chemistry somewhat unclear and subject to question. The role played by the degradation products of naphthenic acids is obviously important, and an improved understanding of naphthenic acid corrosion mechanisms will provide a good starting point for a similar analytical approach to be applied to studying related corrosion mechanisms involving carbon dioxide (released by the decarboxylation reaction). In the case of carbon dioxide, it is assumed that the weak carbonic acid (H_2CO_3) either acts as a reservoir of hydrogen ions (giving rise to the *buffering effect*) and/or can be reduced directly at the steel surface (Hurlen and Gunvaldsen, 1984; Linter and Burstein, 1999; Remita et al., 2008).

Other species in crude oil add complexity to the chemistry of naphthenic acid corrosion. Not all naphthenic acid species in petroleum as determined by the TAN test (Chapter 1) are derivatives of carboxylic acids ($-COOH$), and some of the acidic species (such as phenol derivatives, i.e., derivatives of C_6H_5OH) are resistant to high temperatures. For example, acidic species appear in the vacuum residue after having been subjected to the inlet temperatures of an atmospheric distillation tower and a vacuum distillation tower (Speight and Francisco, 1990; Speight, 2014a). In addition, for the acid species that are volatile, naphthenic acids are most active within the boiling range of the acidic constituents and the most severe corrosion generally occurs at the time of (and shortly thereafter) the condensation of the naphthenic acids from the vapor phase back to the liquid phase.

In addition, petroleum products (especially distillation fractions) are also corrosive. For example, as a general rule, crude oils with an acid number greater than 0.3 to 0.5 mg KOH per gram of crude oil (Chapter 1) and refined crude oil fractions with a TAN higher than 1.5 mg KOH per gram of crude oil fraction have been generally considered to be potentially corrosive due to the presence of naphthenic acid species (Piehl, 1987; Kane and Cayard, 2002). The difference in these two values comes from the concentration of naphthenic acid in specific product fractions produced during the refining process. However, such simple general rules do not always indicate which hydrocarbon fractions and in what locations in the process the concentration of the corrosive acids will occur (Speight and Francisco, 1990). This type of information can potentially result in a better understanding of naphthenic acid corrosivity and help locate potential problem areas in the refinery.

Temperature, metallurgy, TAN, molecular structure, and local flow conditions are known to be important factors affecting naphthenic acid corrosion (Kane and Cayard, 2002; Qu et al., 2007). Metallurgy may not lead to significant differences when it comes to naphthenic acid corrosion. Typically, corrosivity increases significantly with an increase in the TAN above a certain threshold value but at low TAN—under the so-called sulfidation dominated regime—changes in the TAN do not affect naphthenic acid corrosivity, because the corrosion process is controlled by the protectiveness of the iron sulfide scale. The critical TAN value increased for oils with higher total sulfur content (Kanukuntla et al., 2008).

Identification of the chemistry and physics of corrosion allows attempts to be made to prevent corrosion by mitigating the predominant chemical reactions and any associated physical effects (Chapter 5) (Wranglen, 1985; Uhlig and Revie, 1985). Many of the methods for preventing or reducing naphthenic acid corrosion exist, most of them orientated toward removing the naphthenic acids from the feedstock (Chapter 5) or slowing the rates of corrosion (Bradford, 1993; Jones, 1996).

In the former case (i.e., removal of the naphthenic acids from the feedstock), this is the most common approach. In the second case, a series of methods have been developed that are based on depositing a layer of a second material on the surface of a metal structure to impede the structure's contact with the aggressive naphthenic acids (Speight, 2014b). The most prevalent of these is painting, and a wide range of protective paints is now available and included among these surface covering methods are metallic surface treatments, such as chrome, nickel, and galvanized coatings, and inorganic treatments, such as chromates, anodizing coatings, and phosphate coatings. However, another factor that needs to be considered is the high-temperature stability of the coating—such coatings are not always resistant to degradation at the high temperatures associated with many refinery processing units.

As an alternative to using metals that must be protected by one or other of the methods described, there is also the option of using an alloy selected for having a greater resistance to corrosion caused by its surroundings. However, alloys with good resistance in one environment may have poor resistance in another, and their resistance is also likely to vary according to differences in exposure conditions, such as temperature or stress.

Another method of protection uses chemical inhibitors, which are substances added to the liquid medium, again to reduce rates of corrosion (Chapter 5) (Speight, 2014b).

2.3.2 Other Chemical Effects

While corrosion (such as oxidative corrosion, i.e., rust formation) can be generally represented by relatively simple chemical reactions (Speight, 2014b and references cited therein), that is not the complete story. There are various forms of corrosion, some of which involve more complex chemical redox reactions (reduction–oxidation reactions) and many of which do not involve redox reactions (Garverick, 1994). The types of corrosion pertinent to the present section are (1) dry corrosion and (2) wet corrosion which are presented in the following sections.

2.3.2.1 Dry Corrosion

One form of naphthenic acid corrosion is *dry corrosion*, which occurs in the absence of moisture and increases with increasing temperature. At ambient temperature, this form of corrosion occurs on metals that form a rapid thermodynamically stable film, typically in the presence of oxygen. These films are desirable because they are usually free of defects and act as a protective barrier to further corrosive attack of the base metal—metals such as stainless steel, titanium, and chromium develop this type of protective film. Porous and nonadhering films that form spontaneously on nonpassive metals as unalloyed steel are not desirable.

Although the iron sulfide film (formed when hydrogen sulfide is present with naphthenic acids) offers protection against acidic corrosion by retarding the corrosion process, if there is oxidative corrosion the presence of sulfides increases the likelihood of defects in the oxide lattice and thus destroys the protective nature of the natural (oxide) film, which leads to a corroded or pitted surface.

Typically, for naphthenic acids, dry corrosion is not as detrimental as wet corrosion, but it is very sensitive to temperature.

2.3.2.2 Wet Corrosion

Wet corrosion—often subcategorized into *damp corrosion* and *wet corrosion* depending on the environmental conditions—requires moisture in the system and increases in severity with moisture content. When the humidity exceeds a critical value—dependent upon other environmental variables—which is usually in the order of 70% relative humidity, an invisible thin film of moisture will form on the surface of the metal, providing an electrolyte for current transfer. The critical value depends on surface conditions such as cleanliness, corrosion product buildup, and the presence of salts or other contaminants that are hygroscopic and can absorb water at lower relative humidity. Wet corrosion occurs when water occurs on metal surfaces because of sea spray, rain, or other source of moisture.

Crevices or condensation traps also promote the pooling of water and lead to wet atmospheric corrosion even when the flat surfaces of a metal appear to be dry. During wet corrosion, the solubility of the corrosion product can affect the corrosion rate. Typically, when the corrosion product is soluble, the corrosion rate will increase because the dissolved ions increase the conductivity of the electrolyte and thus decrease the internal resistance to current flow, which leads to an increased corrosion rate. Under alternating wet and dry conditions, the formation of an insoluble corrosion product on the surface may increase the corrosion rate during the dry cycle by absorbing moisture and continually wetting the surface of the metal.

Typically, for naphthenic acids, wet corrosion is more detrimental than dry corrosion and may be somewhat less sensitive to temperature.

2.4 SULFIDIC CORROSION

The issue of sulfidic corrosion in refineries is of great importance when considering naphthenic acid corrosion. The presence of sulfur in crude oil can/will enhance the corrosive effects of naphthenic acids in the same oil.

Other than carbon and hydrogen, sulfur is the most abundant element in petroleum. It may be present as elemental sulfur, hydrogen sulfide, mercaptans, sulfides, and polysulfides. Sulfur at a level of 0.2% and greater is known to be corrosive to carbon and low-alloy steels at temperatures from 230°C to 455°C (450−850°F). One reason why sulfidic corrosion (sulfidation) is receiving much more emphasis is the increased processing of low-quality sour (high-sulfur) crude oils, which often contain naphthenic acids.

Sour crude is crude oil with high sulfur content (as opposed to low sulfur content *sweet crude*). Although sour crude oil is available at a lower cost and may be preferable to refineries, low-sulfur (sweet) crude oil is becoming less readily available as the bulk of its supply is exhausted. In sour crude, sulfur is present in the form of mercaptans (RSH), hydrogen sulfide (H_2S), sulfide salts, and a variety of other sulfur-containing constituents (Speight and Ozum, 2002; Hsu and Robinson, 2006; Gary et al., 2007; Speight, 2014a). Sulfur may be present as elemental sulfur (S), hydrogen sulfide (H_2S), mercaptans (RSH), sulfides (SR), and polysulfides (RS_nR).

Sulfur in crude oil at levels greater than 0.2% w/w is been found to be corrosive to carbon steel and also to low-alloy steel at temperatures on the order of 230−455°C (450−850°F) (Backensto et al., 1956; Hucińska, 2006). This is often the result of partial conversion of sulfur constituents to hydrogen sulfide during atmospheric distillation. Usually this form of sulfur corrosion can be handled adequately with steel alloys (5−9% w/w Cr−Mo) unless the crude oil also contains naphthenic acids for which Type 316 or Type 317 stainless steel is preferred. Typically, the most important internal or metallurgical factor to control sulfidic corrosion is the amount of chromium in the steel. The refinery industry relies today in a vast industrial experience on the variables affecting sulfidic corrosion but very little is known on the basic mechanism of attack (Rebak, 2011).

Sulfidic corrosion (sometime referred to as *sulfidation* or *sulfidization*) is differentiated from naphthenic acid corrosion by the corrosion mechanism and the form and structure of the corrosion. While naphthenic acid corrosion is typically characterized as having more localized attack particularly at areas of high velocity and, in some cases, where condensation of concentrated acid vapors can occur in crude distillation units, sulfidic corrosion typically takes the form of a general mass loss or wastage of the exposed surface with the formation of a sulfide corrosion scale.

In addition, the particular forms of sulfur that can participate in this process and the mechanism by which sulfidic corrosion can be understood involves the realization that both sulfur and acid species are present to a varying degree in all crude oils and fractions. In certain limited amounts, sulfur compounds may provide a limited degree of protection from corrosion with the formation of pseudo-passivity sulfide films on the metal surfaces. However, increases in either reactive sulfur species or naphthenic acids to levels beyond their threshold limits for various alloys may accelerate corrosion (Kane and Cayard, 2002).

Sulfidation of metals is subject to (1) exposure time, (2) partial pressure of hydrogen, (3) partial pressure of hydrogen sulfide, (4) temperature, especially to temperatures above 200°C (390°F), and (5) gases containing hydrogen sulfide. Examples of the types of process equipment where sulfidation is a concern are (1) distillation columns, (2) vacuum columns and flashers, (3) coking units, (4) hydrotreater charge furnaces, (5) and sulfur removal plants (gas sweetening plants). Common methods to confirm that sulfidation has occurred are either X-ray diffraction analysis of the surface scale or analysis of the gas composition.

Sulfidation can occur upon exposure of metals to temperatures above approximately 200°C (390°F) in gases containing hydrogen sulfide at extremely low particle pressure. Examples of the types of process equipment where sulfidation is a concern are hydrotreater charge furnaces, crude distilling columns, vacuum flashers, petroleum coking units, and sulfur removal plants (gas sweetening plants). The presence of sulfides confirms sulfidation, which occurs when metals are exposed to gases containing hydrogen sulfide and carbonyl sulfide (COS), and variable process parameters that influence the sulfidation rate are (1) the exposure time, (2) the partial pressure of hydrogen sulfide, and (3) the temperature.

This situation is not expected to improve soon because of the increased processing of low-quality, high-acid, sour (sulfur containing) crude oils. Because it provides a lower feedstock cost (discounted cost) (Chapter 1), such crude oils are preferred by refineries for economic reasons. Furthermore, sweet (low sulfur, low acid) crude oil is becoming less readily available as the bulk of its supply is exhausted. In sour crude, sulfur is present in the form of mercaptans, hydrogen sulfide (H_2S), sulfide salts, and a variety of other sulfur-containing constituents (Hsu and Robinson, 2006; Gary et al., 2007; Speight, 2014a). Many of these species are reactive and combine with naphthenic acids constituents to cause corrosion, leading to stress cracking and sulfuric acid corrosion in other units throughout the refinery.

Sulfidic corrosion of steels in refineries is a prevalent phenomenon that occurs in oil-containing sulfur species between 230°C and 425°C.

There are several internal and external variables controlling the occurrence of sulfidic corrosion. The most important external factors are temperature, concentration and type of sulfur species, and presence of naphthenic acid. The most important internal or metallurgical factor to control sulfidic corrosion is the amount of chromium in the steel. The refinery industry relies today in a vast industrial experience on the variables affecting sulfidic corrosion but very little is known on the basic mechanism of attack (Rebak, 2011).

Boilers generating steam for use in power generation and process power plants use different type of fuels—varying from coke to heavy oil (Speight, 2013a, 2013b, 2014a). In such cases, the presence of naphthenic acids may not be too alarming but the higher the percentage of sulfur, the higher will be the risk of cold-end corrosion in the boiler. The sulfur in the fuel during combustion gets converted to sulfur dioxide. Depending upon the other impurities present in the fuel and excess air levels, some portion of the sulfur dioxide is converted to sulfur trioxide. Because of the presence of moisture in the flue gas (due to moisture in fuel and air), sulfur dioxide and trioxide forms sulfuric acid. These acids condense over the range $115-160°C$ ($240-320°F$), depending upon the concentration of sulfur trioxide and water vapor and leading to the formation of the acid species:

$$S + O_2 \rightarrow SO_2$$
$$SO_2 + O_2 \leftrightarrow SO_3$$
$$H_2O + SO_2 \leftrightarrow H_2SO_3$$
$$H_2O + SO_3 \rightarrow H_2SO_4$$

Depending upon the concentration of sulfur trioxide and water vapor, the dew point temperature can vary from approximately $90°C$ to $140°C$ ($195-285°F$).

Condensation of these acids results in metal wastage and boiler tube failure, air preheater corrosion, and flue gas duct corrosion. In order to avoid or reduce the cold-end corrosion, the gas temperature leaving the heat transfer surface in boiler is kept at approximately $150°C$ ($300°F$) but within the range $120-155°C$ ($250-310°F$). It is very important that the metal temperature of the tubes is always kept above the condensation temperature. It may be noted that the metal temperature of the tubes is governed by the medium temperature of the fluid inside the tubes. This makes it necessary to preheat water to at least $150°C$

(300°F) before it enters the economizer surface. In the case of an air preheater, two methods are used to increase the metal temperature. One is an air bypass for air preheater, and the second is using a steam coil air preheater to increase the air temperature entering the air preheater.

The amount of sulfur trioxide (SO_3) produced in boiler flue gas increases with an increase of excess air, gas temperature, residence time available, the amount of catalysts such as vanadium pentoxide (V_2O_5), nickel (Ni), ferric oxide (Fe_2O_3), and the sulfur level in crude oil or the boiler fuel. In addition, and depending upon the concentration of the sulfur trioxide and water vapor concentration, the dew point temperature can vary from approximately 90°C to 140°C (195–285°F)—this is the temperature below which water vapor in air at constant barometric pressure condenses into liquid water at the same rate at which it evaporates. The condensed water (dew) forms on a solid (metal) surface and is acidic.

In order to avoid or reduce the cold-end corrosion, the gas temperature leaving the heat transfer surface in the boiler is kept at approximately 150°C (300°F) but within the range 120–155°C (250–310°F). It is very important that the metal temperature of the tubes is always kept above the condensation (dew point) temperature. However, the temperature of the metal tubes is governed by the temperature of the fluid inside the tubes—this requires that the water is preheated to at least 150°C (300°F) before it enters boiler tubes. In the case of an air preheater, two methods are used to increase the metal temperature: (1) an air bypass for the air preheater and (2) use of a steam coil air preheater to increase the air temperature entering the air preheater.

2.5 PHYSICAL EFFECTS

Naphthenic corrosion is an aggressive form of local corrosion related to the processing of acidic crude oil (Chapter 1). The typical naphthenic corrosion is observed in the temperature interval 200–400°C (390–750°F). Generally, the corrosion rate increases thrice per each 55°C (99°F) increase of temperature over this temperature range, and it has been suggested that the corrosion is limited within the zones where the condensing acids could contact with metal surface (e.g., lower part of the plates), that is, where the protective layer of hydrocarbons diluting the acids is missing (Petkova et al., 2009).

2.5.1 Effect of Process Parameters

Currently, there are indications that both naphthenic acid corrosion and sulfidic corrosion can be accelerated by velocity of the flowing process environment or by local turbulence (Kane and Cayard, 2002). The wall shear stress produced by the flowing media contributes an added mechanical means to remove the normally protective sulfide films. The wall shear stress is proportional to velocity but also takes into account the physical properties of the flowing media. These properties include density and viscosity of medium (or media) which is, in turn, affected by the degree of vaporization and temperature. This program placed a strong emphasis on quantifying the mechanical forces produced by the flow in terms of wall shear stress, which can act on the surface of operating equipment. This, in turn, is reflected not only in the type of the reactor but also in the process parameters. Not surprisingly, corrosion is reactor dependent.

Although naphthenic acid and sulfidic corrosion are often associated and interact according to the complexity of the process, the two mechanisms can act independently with one dominating the corrosion behavior. Both can interact with fluid velocity, which also invokes the concepts of flow/turbulence and induced wall shear stress.

The sections of the process susceptible to corrosion include preheat exchanger (HCl and H_2S), preheat furnace and bottoms exchanger (H_2S and sulfur compounds), atmospheric tower and vacuum furnace (H_2S, sulfur compounds, and organic acids), vacuum tower (H_2S and organic acids), and overhead (H_2S, HCl, and water). Where sour crudes are processed, severe corrosion can occur in furnace tubing and in both atmospheric and vacuum towers where metal temperatures exceed 450° F. Wet H_2S also will cause cracks in steel. When processing high-nitrogen crudes, nitrogen oxides can form in the flue gases of furnaces. Nitrogen oxides are corrosive to steel when cooled to low temperatures in the presence of water.

However, it is not always possible to correlate corrosion rates for particular crude oils from refinery to refinery (even from a process unit to the process unit) due to the differences in equipment design, operating temperatures, flow velocities, and other crudes present, which may provide a natural passivating effect to the system. Thus, there are several important variables to consider while performing an estimate of the potential for corrosion in a refinery; these variables are stream analysis, temperature, velocity, metallurgy, and flow regimes. Every aspect of the process must be analyzed before the best mitigation strategies (Chapter 5) can be developed.

It is also advisable to conduct such testing on the anticipated blends that could be encountered to ensure the contributions of other crudes to TAN and naphthenic acid titration number. For example, North Sea Captain crude has a whole crude TAN of 2.5 and a naphthenic acid titration number of 2.2, yet the lowest boiling fraction lightest has a TAN on the order of 0.25 mg KOH/g sample and a TAN of 0.3 mg KOH/g sample. The higher boiling fractions have increased TANs increasing to high TAN and naphthenic acid titration levels on the order of 3.8 and 2.6 mg KOH/g sample, respectively, in the fraction boiling at 390−480°C (735−895°F).

2.5.2 Effect of Temperature

Wherever the location, many refineries share similar problems, such as: (1) aging equipment, (2) high costs of replacement, and (3) the need to produce more efficiently while being increasingly concerned with issues of safety and reliability. For equipment operating at various temperatures, especially at high temperature, there are many different mechanisms of corrosion damage, some of which are interactive. In addition, the rate of accumulation of the damage is not always easy to predict, especially when high temperature plays a role in the corrosion process.

When metal is exposed to an oxidizing gas at elevated temperature, corrosion can occur by direct reaction with the gas without the need for the presence of a liquid phase (electrolyte). This type of corrosion is referred to as *high-temperature oxidation, scaling,* or *tarnishing* and increases substantially with temperature. In the refinery and gas processing plant scenario, the rate of most reactions leading to corrosion, as with the rate of chemical reactions in general, increases with increasing temperature, approximating to a doubling of the reaction rate for each 10°C (18°F) rise in temperature whether the corrosion process involves dissolution leading to general attack or to a more localized form such as cracking. In general, therefore, lower temperatures will be more beneficial, but there are exceptions.

High-temperature crude corrosivity of distillation units (due to the presence of naphthenic acids in the feedstock) is a major concern of the refining industry (Craig, 1996; Wang et al., 2011a). Naphthenic acid corrosion occurs primarily in high-velocity areas of crude distillation units in the 220–400°C (430–750°F) temperature range. No corrosion damage is usually found at temperatures greater than 400°C (750°F), probably due to the decomposition of naphthenic acids or protection of the metal from the coke formed on the metal surface.

The presence of naphthenic acid and sulfur compounds considerably increases corrosion in the high-temperature parts of the distillation units, and therefore, equipment failures have become a critical safety and reliability issue. In addition to acidic corrosion, the presence of naphthenic acids may increase the severity of sulfidic corrosion at high temperatures, especially in furnaces and transfer lines. The presence of naphthenic acids can disrupt the sulfide film, thereby promoting sulfidic corrosion on alloys (such as 12 Cr and higher alloys) that would normally be expected to resist this form of attack. In some cases, such as in side-cut piping, the metal sulfide film produced from hydrogen sulfide is believed to offer some degree of protection from naphthenic acid corrosion.

In the refining and gas processing industries, the influence of factors (such as temperature, TAN, fluid velocity, and material) tend to work simultaneously and cumulatively, and in order to obtain credible (low) rates of corrosion, comprehensive analysis and estimation of the propensity for corrosion should be performed on the basis of consideration of temperature and other relevant factors.

In summary, the relationship between temperature and corrosion rate at a constant TAN generally exhibits a linear relationship. Naphthenic acid corrosion is normally not a concern much below 200°C (390°F). As temperature increases, the corrosion rate may increase until the temperature is sufficiently high to decompose the complex higher molecular weight naphthenic acids to lower molecular weight organic acids. The differential between TAN and the naphthenic acid titration number then begins to widen with the naphthenic acid titration number decreasing—this phenomenon usually occurs at temperatures in excess of 420°C (790°F), where the decomposition of the organic species in petroleum occurs at an ever-increasing rate (Speight, 2014a).

High-temperature corrosion is not limited to only naphthenic acid corrosion in the distillation units and can also occur by carburization or sulfidation. Carburization takes place in carbon-rich atmospheres such as in reformer or other (high-temperature) furnaces, and the surface layer of the alloy can become brittle, leading to the formation of cracks, particularly when there are severe or cyclic temperature changes, which can greatly reduce the strength of the component. Sulfidation can be a serious problem in nickel-based super alloys and austenitic stainless steels, with sulfides also forming on grain boundaries and then being progressively oxidized and causing embrittlement in the alloy.

Oxidation is the most commonly encountered form of high-temperature corrosion but may not always be detrimental. In fact, most corrosion- and heat-resistant alloys rely on the formation of an oxide film to provide corrosion resistance. Chromium oxide (chromia, Cr_2O_3) is the most common of such films. However, as the temperature is increased, the rate of oxidation increases and becomes deleterious. Increased chromium content is the most common way of mitigating or at least improving the oxidation resistance of alloys.

While minimization of corrosion in alloys for high-temperature applications depends on the formation of a protective oxide scale, for alloys with very high strength properties at high temperature, a protective coating may need to be applied. The oxides that are generally used to provide protective layers are chromia (Cr_2O_3) and alumina (Al_2O_3). Corrosion protection usually breaks down through mechanical failure of the protective layer, which involves spalling of the oxide (production of flakes produced by a variety of mechanisms, including as a result of projectile impact, corrosion, weathering, or cavitation) as a result of thermal cycling or from erosion or impact.

Thus, naphthenic acid corrosion and high-temperature crude corrosivity in general become a reliability issue in refinery distillation units (Chapter 5). Refinery corrosion occurs at temperatures between 220°C (430°F) and 400°C (750°F). In this temperature range, naphthenic (organic) acids (RCOOH) reach their boiling points and condense on metal surfaces, removing iron [Fe] and eventually causing pits. There may also be decarboxylation and the carbon dioxide that is formed also has the potential to cause corrosion. However, corrosivity does not always correlate with TAN (Derungs, 1956; Zetlmeisl et al., 2000; Messer et al., 2004). Thus, there is the possibility that other organic acid species in some crude oil contribute to corrosion.

Reactive sulfur content of the various side-cut oils also requires investigation as this can lead to strategies for inhibiting the possibility of naphthenic acid attack (with possible decreased corrosion rate). Higher sulfur level is generally considered to be beneficial for the inhibition of naphthenic acid attack. However, increased concentrations of reactive sulfur may trigger high-temperature sulfur-based corrosion. It is important to note that traditional metallurgies resistant to high-temperature sulfidic corrosion are not very resistant to high-temperature naphthenic acid corrosion attack.

The combined presence of naphthenic acid and sulfur compounds considerably increases corrosion in the high-temperature parts of the distillation units. Sulfur-containing compounds decompose to form hydrogen sulfide (H_2S), where iron removal causes general corrosion but can form protective films. Acidic complements of the feedstocks and hydrogen sulfide are often chemically complementary in terms of corrosivity:

$$Fe + 2RCOOH \rightarrow Fe(RCOO)_2(oil\ soluble) + H_2$$

In fact, the ability to predict corrosion behavior has been difficult, and the accepted chemistry that organic acid corrosion, that corrosivity correlates with TAN, is inadequate for predicting refinery corrosion due to its assumption that all acid molecules are equally corrosive regardless of their composition and structure (Zetlmeisl et al., 2000).

Furthermore, the difference in process conditions, materials of construction, and blend processed in each refinery and especially the frequent variation in crude diet increases the problem of correlating corrosion of a unit to a certain type of crude oil. In addition, crude oil composition from the same field can change with time. When steam flooding or other recovery methods begin in an oil field, the specific gravity and the organic and sulfur content of the crude may change. For example, fire flooding, when used in some fields, tends to increase the naphthenic acid content.

There are at least three mechanisms of naphthenic acid corrosion. Each one is predominant in specific areas of the distillation unit (Chapter 3).

CHAPTER 3

Corrosion by High Acid Crude Oil

3.1 INTRODUCTION

The trend in crude oil supplies by refineries is towards heavier, lower quality feedstocks—the ratio of heavy crude oil (API 10−20°) in the total crude oil slate continues to rise, often in an apparent accelerated pace (Speight, 2014a). Thus, the oil refining industry must increasingly face the challenges presented by heavy, low quality crudes, cleaner production standards, and the demand for cleaner fuels and high-value petrochemical products.

Opportunity crude oil generally refers to crude oils with relatively high metals and sulfur content, and a high total acid number (TAN) and density (see Chapter 1). High acid content in heavy crude oil is the typical opportunity crude oil and the following properties are general: (i) fewer low boiling components, (ii) high density, (iii) low viscosity, (iv) high metal contents, (v) high asphaltene content, and (vi) high acid number, which give rise to equipment corrosion and severe problems with product quality and environmental protection.

The growing variety of discounted opportunity crudes on the market usually contains one or more risks for the purchaser, such as high naphthenic acid content. As the availability and volume of highly naphthenic crudes processed increase, the risk of experiencing high temperature corrosion on refinery equipment must be considered. In fact, many opportunity crudes are known to contain naphthenic acids (NAP), which can cause corrosion in high temperature regions within the refinery, normally around the crude and vacuum towers. These opportunity crudes, also known as *high acid crude oils*, are often discounted due to the added risks associated with processing these crudes. Because of the economic advantages, many refiners are looking increasingly at processing high levels of naphthenic crude oils in their crude slates.

NAP have been identified as the main corrosive species in acidic crudes although they represent <4% (w/w) of the crude oil (see Chapter 1) (Dzidic et al., 1988; Fan, 1991; Hsu et al., 2000; Laredo et al., 2004).

Naphthenic acids in petroleum (NAP) are corrosive at high temperatures and the lower temperature limit when petroleum naphthenic acids become corrosive is 220°C (430°F) (Derungs, 1956). The corrosive effects of petroleum naphthenic acids are intense at a temperature range between 220°C (430°F) and 400°C (750°F) (Heller et al., 1963; Slater et al., 1974; Gutzeit, 1976a,b, 1977; Morrison et al., 1992; Babaian-Kibala et al., 1993a,b; Craig, 1995, 1996; Slavcheva et al., 1998; Turnbull et al., 1998). There are also two methods developed by UOP (UOP Method 565-92, UOP Method 587-92) in which the sulfur is first removed before determining the acid number. This negates any effects of acid sulfur compounds on the TAN.

The TAN has been used to distinguish the acidity of crude oil: acidic crude oil has an acid number >0.5 mg KOH/g and high acid crude typically has an acid number >1.0 mg KOH/g, although this number is subject to other factors that render it of questionable value (see Chapter 1). For high acid crude, there are two main types that are designated according to sulfur content: (i) high TAN low sulfur heavy crude and (ii) high TAN high sulfur heavy crude. The latter is produced in Venezuela and California, while high acid low sulfur heavy crudes are more common and there are oil fields in four continents producing this type of crude, which is processed by refineries in Europe and the Gulf of Mexico. In the Americas, the high acid low sulfur heavy crudes are mainly represented by the Marlim crude oil (Campos Basin, Brazil) and are processed in several continents. In Africa, the high acid low sulfur heavy crudes are mainly represented by West African Kuito crude oil and Sultan crude oil, which are processed in refineries in the United States and Asia. In Asia, high acid low sulfur heavy crudes are mainly represented by crudes such as Penglai crude oil (Bohai Bay, China), which is processed mainly in China.

3.2 PROCESS EFFECTS

Because of their economic attractiveness, there is widespread interest in processing corrosive crudes, particularly those with high TAN. Many refiners have conducted studies to assess the amount of corrosive crude that can be processed with their existing equipment (i.e., without changing its metallurgy). Also, they must often determine the extent to which their equipment and piping construction materials must be upgraded in order to be able to process the desired crude.

However, the properties of crude oils rich in naphthenic acid species vary and no single approach is best for all sites—the answer is different for each site. For example, Captain, Alba, Gryphon, Harding, and Heidrun are all considered opportunity crudes from the North Sea region. The specific gravity of these crudes ranges from 0.88 to 0.94 and acid numbers range from 2.0 to 4.1. The Gryphon has a sulfur content of 0.4 wt% while Alba crude oil contains 1.3% (w/w) sulfur. These variations allow refiners to find high acid crude oils that are suitable for their product profile. In addition, North Sea Captain crude oil has a whole crude TAN of 2.5 and a naphthenic acid titration number of 2.2, yet the lowest boiling fraction (lightest cut) only has a TAN of 0.25 and a naphthenic acid titration number of 0.3. As higher boiling fractions (heavier cuts) are tested, the TAN and the naphthenic acid levels increase to a high TAN of 3.8 and naphthenic acid titration number of 2.6 in the 390–480°C (735–900°F) cut.

NAP cause several problems during refining and from an operational point of view, producing and processing highly acidic crudes involve a number of challenges. Corrosion by NAP, especially in the high temperature parts of the distillation units, is a major concern to the refining industry (Piehl, 1988; Babaian-Kibala et al., 1993a,b; Slavcheva et al., 1998, 1999). In addition and prior to entering the refinery, there are also problems that arise during production of high acid crudes—the amphiphilic NAP may also accumulate at interfaces and stabilize water-in-oil emulsions causing enhanced separation problems (Pathak and Kumar, 1995; Acevedo et al., 1999; Goldszal et al., 2002; Ese and Kilpatrick, 2004). Pressure drop during fluid transportation from the reservoir to the topside leads to release of carbon dioxide and to a subsequent increase in the pH of the coproduced water, which brings about a higher degree of dissociation of NAP at the oil–water interface. The dissociated moieties may thus react with metal cations in the water to form metal naphthenates. Due to low interfacial affinity and low solubility in water, especially when multivalent cations are involved, the naphthenates will precipitate and start to agglomerate in the oil phase. As the density of the precipitate lies between that of oil and water, it will gradually settle and accumulate at the oil–water interface and further adhere to process unit surfaces. Deposition of metal naphthenates (see Chapter 1) is a problem predominantly in topside facilities like oil–water separators and desalters, and may lead to difficulties in desalting leading to costly shutdowns.

Many refiners have opted to install high alloy equipment and piping systems that are resistant to corrosion by naphthenic acids. While effective, this approach is only economically attractive if a secure long-term supply of corrosive crude is available at attractive prices. Money spent on alloy material can only be recovered by processing corrosive crudes at a good margin. Experience has shown that the most effective approach is to establish a series of limits on TAN (or neutralization number) that are related to progressive corrosion control actions. Preferred corrosion control actions include the following: (i) use of a naphthenic acid corrosion (NAC) inhibitor, (ii) focused inspection and corrosion monitoring, and (iii) limited alloy upgrading.

The advantage of this approach is that inhibitor injection can be discontinued and monitoring can be reduced when high TAN crude is not available. This reduces the total cost for the program.

Crude oil corrosivity is a function of the TAN and sulfur content, amongst other parameters. Crude oils that have a high TAN and low sulfur are particularly corrosive. It is not always possible to develop a series of operating envelopes in terms of TAN and sulfur (i.e., corrosive sulfur compounds). For each operating envelope, a specific corrosion control action is required. The economically best option is to run to the limit of a chosen corrosion control level. This approach maximizes the amount of corrosive crude oil that can be processed for a specific level of corrosion control.

Experience has shown that naphthenic acid crude oil feedstocks generally first affect the vacuum transfer line and vacuum gas oil (VGO) side stream (Figure 3.1). As the feed TAN is further increased, corrosion will affect the atmospheric transfer line and heavy atmospheric gas oil (HAGO) circuits and the bottoms of both towers.

NAC is complicated not least because of the complexity of the mixture of naphthenic acid compounds found in crude oils from the various sources. Sometimes, the distribution of NAP is grouped according to their boiling point, with an implication that NAP with different boiling points lead to different corrosivity. Concurrent presence of both sulfur containing compounds and NAP in crudes introduces a new level of complexity into the study of the corrosion mechanisms. Sulfidation corrosion leads to formation of a sulfide scale, which

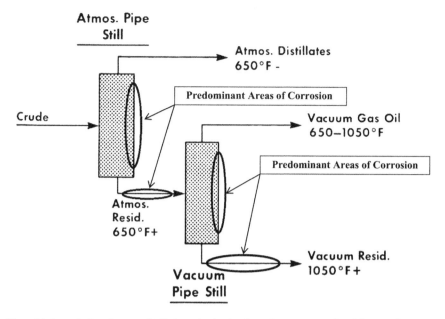

Figure 3.1 Atmospheric and vacuum distillation units showing the predominant areas of naphthenic acid corrosion.

provides some degree of protection against NAC; however, this subject has been given relatively little systematic attention in the past.

In order to process high acid crude oils, it is necessary to use a series of steps based on reliable background information which will provide an approach to evaluating processing options for high acid number corrosive crude oils and will allow development of corrosion mitigation methodologies for high acid corrosive crude oils.

For example, data on the materials of construction and the condition of equipment and piping systems should be collected and analyzed—the data should include information about fired heaters, transfer lines, lower tower, lower side streams, and bottoms of the atmospheric distillation unit and the vacuum distillation unit. Also, it is necessary to summarize any current corrosion control programs and their results for atmospheric distillation overhead system as well as the performance of the desalting operation. Moreover, the records of the past crude slate and performance properties of each crude oil (especially the properties of crude oil blends) should be summarized and used as back-up data to be evaluated as may be applied to future crude slates.

It is also necessary to determine the predicted equipment life for both the current crude slate and future crude feedstock options, which must also include any limits on the lifetime of equipment and piping that would be affected by corrosion. As part of this exercise, there should also be an investigation of the TANs and sulfur content that would affect equipment and piping performance that will permit acceptable run lengths. Actions to enhance equipment and piping performance should also be identified and any incremental increases in the TAN of the feedstock should be identified and the influence on equipment and piping estimated.

A most important aspect of this work is to develop a monitoring and inspection program which is practical and which will allow the predicted limits to be optimized. As part of this monitoring program, it is essential to define the baseline monitoring to be carried out before an increase in the TAN of the feedstock as well as to establish review intervals for the overall program. As part of corrosion mitigation, determine the most attractive combination of inspection, inhibition, and limited alloy upgrading for several levels of TAN of the feedstock. Some high acid crude oils perform poorly in the desalting operation and the result is increased rates of fouling in preheat equipment. Monitoring programs to assess the impact of these issues must also be developed as part of the study.

In summary, processing high acid crude oils is subject to the following criteria as they relate to corrosivity: (i) temperature, (ii) NAP concentrate in fractions boiling above 230°C (450°F), (iii) the highest concentration of NAP is typically found in the 315−425°C (600−800°F) boiling range, (iv) the lowest temperature where attack occurs is approximately 200°C (390°F), (v) lower molecular weight acids—lower boiling acids such as formic acid and acetic acid—occur at water condensing locations, (vi) at low velocity, turbulence caused by boiling and condensing causes attack, and (vii) at high velocity, rapid corrosion can occur. These criteria are reasonably well defined for conventional crude oil but are somewhat less well defined for heavy crude oil and tar sand bitumen.

3.3 CORROSION OF REFINERY EQUIPMENT

The corrosion of refinery equipment during oil distillation was observed during the early part of the twentieth century (Jayaraman et al., 1986). As the oil refining industry evolved toward the modern industry, the problem of corrosion increased and became a serious aspect of equipment performance that needed constant attention (Lewis et al., 1999; Johnson et al., 2003a,b). Experience has shown that the main sites of corrosion attack are the components of oil pumping and refinery transfer lines, such as pipelines, valves and gates, heat exchangers, pipe stills, bubble sections, hydrocarbon stock feeders, and fractionating tower reflux units. Accidents related to corrosion processes resulted in equipment outage and economic losses, thus making it necessary to investigate the causes of this type of corrosion. Petroleum naphthenic acids (PNA) present in crude oils produced in many world regions are considered now to be the main factor responsible for the corrosion problem (Gutzeit, 1976a,b, 1977; Jayaraman et al., 1986; Babaian-Kibala et al., 1993a,b; Turnbull et al., 1994; Slavcheva et al., 1998, 1999).

As oil processing temperatures increase, NAP exhibit significant corrosivity, inducing the specific corrosion type termed *NAC*. The nature of this type of corrosion and the factors controlling it are still incompletely understood (Zetlmeisl, 1996). Therefore, despite the numerous efforts to prevent or mitigate corrosion made in the last several decades, the problem of corrosion remains acute and ever-present. It is noted that there are two reasons behind this situation: the extreme complexity and interplay of the factors affecting the corrosion and erosion–corrosion processes. These factors are: (i) different corrosivity of crude oils, depending on the total add number (TAN) of the crudes, the activity of PNA and their distribution over boiling points and decomposition temperatures, and the presence of additional corrosion-active compounds, such as organic sulfur compounds and chlorides in crude; (ii) the variable oil refining parameters, such as the hydrocarbon feedstock flow rate, the extent of oil evaporation, and the processing temperature; and (iii) the susceptibility of metal equipment to corrosion. The second reason is the lack of laboratory units for effective restored state experiments mimicking the actual high temperature and high feedstock flow rate conditions. The situation in oil refining has been aggravated in recent years by the progressively increasing involvement of crude oils with a high acid content (Kane and Cayard, 1999).

The issues of NAC have been the subject of many investigations throughout the world; countries such as China, India, Venezuela, Western Europe, Russia, the United States, and Middle Eastern countries have put considerable effort into understanding the mechanism of corrosion with the goal of mitigating damage to refinery equipment (Jayaraman et al., 1986). Earlier, it was believed that this type of corrosion, which is especially intense at high temperatures, is caused by the presence of NAP or sulfur containing constituents of crude oil but researchers failed to distinguish between these factors. Now it is known that the naphthenic acid content is, in principle, related to (i) the TAN of the crude oil system, (ii) the process temperature, and (iii) the flow rate (see Chapters 1 and 2) (Jayaraman et al., 1986; Babaian-Kibala et al., 1993a,b).

However, the TAN is, at best, a rough estimate of the corrosivity of crude oil and it provides investigators only with the general information that oils with a high TAN (0.3 mg KOH/g oil) are already corrosive and does not give accurate information about the extent of the anticipated corrosion. Indeed, there are data that oils with a relatively low TAN are comparable in the corrosivity index with those with a high TAN. Furthermore, crude oil with a high TAN turned out to be less aggressive than could be expected on the basis of the TAN. For example, Indonesian light crude with a TAN <0.5 mg KOH/g and Nigerian crude oil with a TAN 0.3 mg KOH/g and a sulfur content of 0.24% (w/w) causes significant corrosion although the respective TANs do not suggest such a difference in corrosivity (see Chapter 1) (Jayaraman-Kabila et al., 1986; Slavcheva et al., 1999).

Furthermore, it is now generally accepted crude oils having identical TANs differ substantially in corrosivity and, thus, evaluation of crude oil corrosivity in terms of the TAN is insufficient (Turnbull et al., 1994; Slavcheva et al., 1998). A natural explanation offered for these findings is that corrosion is controlled by the type, structure, and functionality of PNA, rather than the TAN value (Qu et al., 2007).

NAC is especially intense at oil distillation temperatures of 220−400°C (430−750°F) and high fluid flow rates. The presence of sulfur promotes naphthenic acid corrosivity owing to the following concurrent reactions (see Chapter 2):

$$Fe + RCOOH \rightarrow Fe(RCOO)_2 + H_2 \qquad (3.1)$$

$$Fe + H_2S \rightarrow FeS + H_2 \qquad (3.2)$$

$$Fe(RCOO)_2 + H_2S \rightarrow FeS + 2RCOOH \qquad (3.3)$$

Reaction (1) illustrates the direct attack of NAP on the steel lining of columns (low-carbon steel) while Reaction (2) is responsible for hydrogen sulfide corrosion. Iron naphthenate as a corrosion product is well soluble in oils, and iron sulfide creates 11 protecting films on the metal surface via Reaction (2). This type of film does not form unless the sulfur content in oil is at least 2–3% (w/w) (Jayaraman et al., 1986). By Reaction (3), dissolved iron naphthenate interacts with hydrogen sulfide to regenerate the naphthenic acid(s) and to produce iron sulfide (FeS), which precipitates in the oil and thereby affecting the quality of the oil.

Investigations of the influence of sulfur compounds on the chemical and physical processes occurring during NAC have shown that the inherent presence of certain sulfur compounds in crude oil can have diametrically opposed consequences of NAC. For example, (i) when the system contains sulfur compounds that produce hydrogen sulfide, a protective iron sulfide (FeS) film is formed on the metal surface and the further corrosion of metal is lessened − if not stopped − depending on the extent of the coverage of the metal by the film, or (ii) if the system contains sulfoxides—oxidation products of the sulfur constituents of crude oil—water forms in the chemical cathode zone and the corrosion processes are intensified (Yépez, 2005).

$$\overset{}{\underset{\text{Sulfide}}{\diagup S \diagdown}} \longrightarrow \overset{\displaystyle \overset{O}{\|}}{\underset{\text{Sulfoxide}}{\diagup S \diagdown}}$$

These observations emphasize that it is important to control the presence of sulfoxides in crude oils because these compounds pose a hazard as promoters of petroleum acid corrosion.

Attempts to draw a correlation between the TAN, sulfur content, and the corrosion activity of crude oil and light and heavy VGOs have shown that there is a good correlation between the TAN and corrosivity only for desalted (demineralized) samples of a 90/10 Isthmus/Maya crude blend (Laredo et al., 2004). For the same fractions, no correlation between the decree of corrosion and the sulfur content was found. The mass spectra of crude samples indicate that the naphthenic acid composition (simple or complex) has no direct effect on the TAN or the corrosivity of tile acids.

It has also been noted (Turnbull et al., 1994; Slavcheva et al., 1998, 1999) that in light of recent studies, the TAN criterion is not useful as a characteristic of corrosivity anymore. In this context, methods for the exact analysis of naphthenic acid structure are becoming of great importance, as well as revealing the correlation between the configuration and functionality of NAP and the corrosivity (Qu et al., 2007). Mass spectrometry, which is widely used for analyzing complex hydrocarbon mixtures, fuel and oil composition (Speight, 2001, 2002, 2014a) and identifying the naphthenic acid structure, is considered to be the most effective and authentic technique. Various hybrid techniques which are versions of mass spectrometry coupled with another method are also in use. The basic difficulty met by researchers in the application of any of these techniques is that of selective isolation of NAP from oil samples, since the mass spectra of the NAP are complex, superimposed, and poorly readable. Researchers also need to consider the three-dimensional electronic structure of the NAP as this most likely will offer more valuable information about the corrosion mechanism.

The corrosivity of oil fractions and the means of its prevention were studied. Kerosene: free of acid components and phosphoric acid esters were tested as corrosion inhibitors. The efficiency of inhibitors was determined on special plates in atmospheric vacuum distillation columns by measuring the electrical resistance of samples. Kerosene showed the highest anticorrosive efficiency at a volume ratio of 1:3. Phosphate inhibitors at a 100 ppm concentration preclude corrosion by almost 100%.

3.3.1 Corrosion by NAP

NAC (see Chapter 1) is a type of high temperature corrosion that occurs in refineries—primarily in atmospheric distillation and vacuum distillation as well as in downstream units that process certain crude oil fractions that contain naphthenic acids (Derungs, 1956; Piehl, 1960, 1988; Gutzeit, 1976a,b). With the increased input of heavy feedstocks (such as heavy oil and tar sand bitumen) (Hopkinson and Penuela, 1997; Messer et al., 2004) that often contains high amounts of acid constituents, other units (such as coking units, fluid catalytic cracking units, and hydrotreating/hydrocracking units), which are often the unit to which the heavy feedstocks are processed directly, i.e., without distillation, are also subject to NAC. The affected materials in the various units are carbon steel, low alloy steels, 300 series stainless steel,

400 series stainless steel, nickel-based alloys, as well as any more recent types of stainless steel.

NAC is a function of several factors: (i) the naphthenic acid content of the feedstock, (ii) temperature, (iii) sulfur content, (iv) feedstock velocity through the reactor, and (v) alloy composition (see Chapters 1 and 2). Generally, the severity of corrosion increases with increasing acidity of the feedstock (remembering that several acidic species make up the total acidity of the feedstock).

NAC is typically associated with hot dry hydrocarbon streams that do not contain a free water phase but if the acids have the opportunity to condense in a water phase, wet corrosion (see Chapter 2) can and will ensue. Additionally, the various acidic constituents which comprise the naphthenic acid family can have distinctly different corrosivity and there are no widely accepted methods that have been developed to correlate or predict corrosion rate with the various factors influencing it (see Chapters 1 and 2) (Tebbal et al., 1997; Tebbal, 1999).

Sulfur promotes iron sulfide formation and has an inhibiting effect (due to the formation of an iron sulfide coating on the metal/alloy) on NAC unless the NAP remove protective iron sulfide scales on the surface of metals. Also, following on from the poor predictability of NAC, it can be a particular problem with low sulfur crude oils that have TANs as low as 0.10 mg KOH/g crude oil (Nugent and Dobis, 1998).

While NAC typically occurs in hot crude oil streams at temperatures in excess of above 220°C (430°F), corrosion has been reported as low as 175°C (345°F). The severity of the corrosion has generally increased with temperature up to approximately 400°C (750°F) but has been observed in coking units at temperatures on the order of up to 425°C (795°F). Furthermore, NAP are destroyed by catalytic reactions in downstream fluid catalytic cracking and hydroprocessing and alloys containing increasing amounts of molybdenum (Mo) show improved resistance to NAC—a minimum of 2–2.5% (w/w) Mo is required depending on the TAN of the crude oil (Shargay et al., 2007).

Finally, corrosion is most severe in two phase (liquid and vapor) flow, in areas of high velocity or turbulence, and in distillation towers where hot vapors condense to form a liquid phase.

The predominant units affected by NAC are atmospheric and vacuum heater tubes, atmospheric and vacuum transfer lines, vacuum bottoms piping, atmospheric gas oil circuits, light VGO circuits, and heavy VGO circuits. NAC has also been reported in the light cycle gas oil and heavy cycle gas oil streams on delayed coking units processing high acid crudes (or resids). Piping systems are particularly susceptible in areas of high velocity, turbulence, or change of flow direction, such as pump internals, valves, elbows, tees and reducers, as well as areas of flow disturbance such as weld beads and thermowells. Atmospheric and vacuum tower internals may also be corroded in the flash zones, packing, and internals where high acid streams condense or high-velocity droplets impinge.

NAC is characterized by localized corrosion and pitting corrosion (a localized form of corrosion by which cavities or holes are produced in the material) (see Chapter 2) in high fluid velocity areas. In low velocity condensing conditions, many alloys including carbon steel, low alloy steels, and 400 Series SS may show uniform loss in thickness and/or pitting.

Prevention and mitigation of such corrosion: the options are to change or blend crudes, upgrade the reactor/unit metallurgy, utilize chemical inhibitors or some combination thereof (see Chapter 5). For severe corrosion conditions, Type 317L stainless steel or other alloys with higher molybdenum content may be required.

Finally, sulfidation is a competing and complimentary mechanism which must be considered in most situations with naphthenic acid. In cases where thinning is occurring, it may be difficult to distinguish between naphthenic acid corrosion and sulfidation.

3.3.2 Corrosion by Organic Acids

NAP, as well as organic compounds present in some crude oils, decompose in the atmospheric unit furnace (the furnace heating the crude oil before injection into the atmospheric tower) to form low molecular weight organic acids (RCO_2H, such as acetic acid, CH_3CO_2H) which condense in distillation tower overhead systems. These acids may also result from additives used in upstream operations or used in the desalting unit and may contribute significantly to aqueous corrosion depending on the type and quantity of acids, as well as the presence of other contaminants.

As with NAC, organic acid corrosion is a function of the type and quantity of organic acids, metal temperature, fluid velocity, system pH, and presence of other acids. The low molecular weight organic acids that are formed include formic acid, acetic acid, propionic acid, and butyric acid—the lower molecular weight acids such as formic acid and acetic acid are the most corrosive. These acids are soluble in naphtha and are extracted into the water phase, once the water condenses, and contribute to a reduction of pH. The presence of organic acids contributes to the overall demand for neutralizing chemicals but their effects may be completely masked by the presence of other acids such as hydrogen chloride (HCl), hydrogen sulfide (H_2S), carbonic acid (H_2CO_3), and others. This type of corrosion is typically manifested where relatively noncorrosive conditions exist in an overhead system and there is a sudden increase in low molecular weight organic acids that reduces the pH of the water in the overhead system requiring a potentially unexpected increase in neutralizer demand.

The type and quantity of organic acids formed in the overhead system are crude specific. One source of acid is believed to be the result of thermal decomposition of NAP in the crude may be precursors to light organic acid formation and that processing of higher TAN crude oils may increase organic acid in the overheads but very little published data is available on this subject.

Some of the higher molecular weight organic acids condense above the water dew point in overhead systems but are generally not present in sufficient quantities to cause corrosion. Additives, including low molecular weight organic acids such as acetic acid, are sometimes added to upstream dehydrators or desalters to improve performance and inhibit calcium naphthenate salt deposition (Kapusta et al., 2003, 2004; Lordo et al., 2008). Such acids will vaporize in the atmospheric distillation unit preheater and furnace, and go up the column into the crude tower overhead system. Generally, the lower molecular weight organic acids do not generate the severity of corrosion associated with inorganic acids such as hydrogen chloride.

In fact, acetic acid is recognized as an important factor in mild steel corrosion. Like carbonic acid, acetic acid is a weak acid, which partially dissociates as a function of pH and the solution temperature. Stronger than carbonic acid (CH_3CO_2H: pK_a 4.76 at 25°C/77°F compared to H_2CO_3: pK_a 6.35 at 25°C/77°F), acetic acid is the main source of hydrogen ions when the concentration of each acid is the same. Furthermore, acetic acid enhances the corrosion rate of mild steel by accelerating the cathodic reaction but the actual mechanism of acetic acid reduction at the metal surface is still not conclusively clear. When the reduction of the adsorbed acetic acid molecule occurs at the metal surface, the mechanism is *direct reduction* but if the role of acetic acid is to dissociate near the metal surface to provide additional hydrogen ions and the only cathodic reduction is reduction of hydrogen ions, this mechanism is referred to as a *buffering effect* (see Chapter 2), which appears to be the correct (Tran et al., 2013).

Thus, corrosion by the lower molecular weight acids can affect all grades of carbon steel piping and process equipment in atmospheric tower, vacuum tower, and coker fractionator overhead systems including heat exchangers, towers and drums, which are susceptible to damage where acidic conditions occur. Moreover, corrosion tends to occur where water accumulates or where hydrocarbon flow directs water droplets against metal surfaces.

Corrosion is also sensitive to flow rate and will tend to be more severe in turbulent areas in piping systems, including overhead transfer lines, overhead condensers, separator drums, control valves, pipe elbows and tees, and exchanger tubes.

The corrosive effect of the lower molecular weight acids typically leaves the corroded surface smooth and damage may be difficult to distinguish from corrosion by other acids in the overhead system—it may be mistaken for hydrochloric acid corrosion or carbonic acid corrosion.

Corrosion caused by low molecular weight organic acids in the atmospheric distillation unit overhead systems can be minimized through the injection of a chemical neutralizing additive (Rue and Naeger, 2007; Braden et al., 2007). However, problems may arise when frequent changes in crude blends lead to changes in neutralizer demand. The TAN of the crude oils being processed can be used as an initial guide to setting the neutralizer by anticipating an increase in the acid concentration in the overhead system. However, after a new (different) crude oil is processed, a review of analyses of water samples from the boot of the overhead separator drum can be used to determine how much light organic acid reaches the overhead system to optimize future additions.

Finally, upgrading to corrosion-resistant alloys will prevent organic acid corrosion but the selection of suitable materials should account for other potential damage mechanisms in the overhead system (Speight, 2014b).

3.3.3 Corrosion by Phenol Derivatives

Corrosion of carbon steel can occur in refineries where phenol or phenol derivatives are present in feedstocks. The materials affected by, and subject to corrosion by, phenol and its derivatives include carbon steel, stainless steel 304L, stainless steel 316L, and the steel alloy C276.

Critical factors in the corrosion process include (i) temperature, (ii) water content of the feedstock, and (iii) fluid velocity. Corrosion is usually minimal where the temperature is below 120°C (250°F) but corrosion must be anticipated in any unit where phenol and its derivatives are separated by vaporization and high fluid velocity may promote localized corrosion. In conjunction with corrosion by phenols, sulfur and organic acids may lead to naphthenic acid attack and sulfidation in the hotter parts of the reactor system. The corrosion will be in the form of general or localized corrosion of carbon steel and localized loss in thickness due to erosion—corrosion may occur—erosion—corrosion and/or condensation corrosion may be observed in tower overhead circuits.

Phenol corrosion is best prevented through proper materials selection and control of phenol solvent chemistry. Overhead piping circuits should be designed for a maximum velocity of 30 ft/s in the recovery section and recovery tower overhead temperatures should be maintained to at least 17°C (30°F) above the dew point. Type 316L stainless steel may be used in the top of the dry tower, phenol flash tower and various condenser shells and separator drums that handle phenol-containing water.

3.4 INTERACTION OF ACIDS WITH REFINERY EQUIPMENT

NAP in crude oil cause corrosion which often occurs in the same places as high temperature sulfur attack such as heater tube outlets, transfer lines, column flash zones, and pumps (Shalaby, 2005 and references cited therein). Furthermore, NAP alone or in combination with other organic acids, such as phenols, can cause corrosion at temperatures as low as 65°C (150°F) and as high as 420°C (790°F)

(Gorbaty et al., 2001; Kittrell, 2006). Crude oils with a TAN higher than 0.5 and crude oil fractions with a TAN higher than 1.5 are considered to be potentially corrosive between the temperature of 230–400°C (450–750°F).

Corrosion by NAP typically has a localized pattern, particularly at areas of high velocity and, in some cases, where condensation of concentrated acid vapors can occur in crude distillation units. The upper head of distillation columns at refineries is in general strongly affected by corrosion. The corrosion reduces the service life of the equipment and creates economic problems for oil refiners. The basic classes of corrosion inhibitors used to prevent the corrosion of fractionation column heads, the criteria for selection of corrosion inhibitors for other industrial units, and the basic aspects of design of the injection system for these inhibitors are considered.. Damage is in the form of unexpected high corrosion rates on alloys that would normally be expected to resist sulfidic corrosion (particularly steels with more than 9% Cr). In some cases, even very highly alloyed materials (i.e., 12% Cr, type 316 stainless steel (SS) and type 317 SS), and in severe cases even 6% Mo stainless steel have been found to exhibit sensitivity to corrosion under these conditions.

The corrosion reaction processes involve the formation of iron naphthenates:

$$\begin{aligned} Fe + 2RCOOH &\rightarrow Fe(RCOO)_2 + H_2 \\ Fe(RCOO)_2 + H_2S &\rightarrow FeS + 2RCOOH \end{aligned} \tag{3.4}$$

The iron naphthenates are soluble in oil and the surface is relatively film free. In the presence of hydrogen sulfide, a sulfide film is formed which can offer some protection depending on the acid concentration. If the sulfur containing compounds are reduced to hydrogen sulfide, the formation of a potentially protective layer of iron sulfide occurs on the unit walls and corrosion is reduced (Kane and Cayard, 2002; Yépez, 2005). When the reduction product is water, coming from the reduction of sulfoxides, the NAC is enhanced (Yépez, 2005).

Thermal decarboxylation can occur during the distillation process (during which the temperature of the crude oil in the distillation column can be as high as 400°C):

$$R - CO_2H \rightarrow R - H + CO_2$$

However, not all acidic species in petroleum are derivatives of carboxylic acids (−COOH) and some of the acidic species are resistant to high temperatures. For example, acidic species appear in the vacuum residue after having been subjected to the inlet temperatures of an atmospheric distillation tower and a vacuum distillation tower (Speight and Francisco, 1990). In addition, for the acid species that are volatile, NAP are most active at their boiling point and the most severe corrosion generally occurs on condensation from the vapor phase back to the liquid phase.

NAC is one of the serious long known problems in the petroleum refining industry (Derungs, 1956; Piehl, 1960; Gutzeit, 1976a,b; Piehl, 1988; Slavcheva et al., 1998, 1999; Qu et al., 2005, 2006). The rates of NAC increase with the temperature rising and are also influenced obviously by TAN, that is, the corrosion rates increase with the TAN increasing. Under the condition of low velocity of naphthenic acid, the influences of fluid velocity on corrosion rate can be detected, especially in the high temperature conditions (Wang et al., 2011). In petroleum refining industry, influence factors (such as temperature, TAN, fluid velocity, and material) work simultaneously, so in order to obtain the credible corrosion rate, the comprehensive analysis and estimation should be done on the basis of considering the combining temperature with other factors.

High temperature naphthenic acid corrosion mainly occurs at temperatures above 200°C (390°F), in particular above 220°C (430°F), affecting equipment that has the closest contact with naphthenic acid. The most seriously affected component is the vacuum distillation column system of the atmospheric and vacuum distillation system, including the vacuum heater, transfer line, vacuum draw three, vacuum draw four, vacuum tower feeding line, and internal structures. High temperature naphthenic acid corrosion includes four main steps: (i) naphthenic acid molecules are transferred to a metal surface, (ii) the molecules become adsorbed onto the metal surface, (iii) the molecules react with surface active centers, and (iv) corroded materials are desorbed.

NAC is heavily influenced by temperature—typically corrosion at low temperatures is not significant but, in the boiling state especially in a high temperature, anhydrous environment corrosion is most significant. Most high temperature naphthenic acid corrosion occurs in the liquid phase, but if naphthenic acid is condensed in the gas phase then gas phase corrosion may occur and the extent of corrosion will be influenced by acid value.

Isolated deep pits in partially passivated areas and/or impingement attack in essentially passivation free areas are typical of NAC. Damage is in the form of unexpected high corrosion rates on alloys that would normally be expected to resist sulfidic corrosion. In many cases, even very highly alloyed materials (i.e., 12 Cr, AISI types 316 and 317) have been found to exhibit sensitivity to corrosion under these conditions. NAC is differentiated from sulfidic corrosion by the nature of the corrosion (pitting and impingement) and by its severe attack at high velocities in crude distillation units. Crude feedstock heaters, furnaces, transfer lines, feed and reflux sections of columns, atmospheric and vacuum columns, heat exchangers, and condensers are among the types of equipment subject to this type of corrosion.

Crude corrosivity is a function of the TAN, sulfur content, and possibly other unidentified factors. Crude oils that have high TANs and low sulfur are particularly corrosive. It is possible to develop a series of operating envelopes in terms of the TAN and sulfur (i.e., corrosive sulfur compounds). For each operating envelope, a specific corrosion control action is required. The economically best option is to run to the limit of a chosen corrosion control level. This type of approach maximizes the amount of corrosive crude that can be processed for a given level of corrosion control.

Experience has shown that naphthenic crudes generally first affect the vacuum transfer line and VGO side stream. As the feed TAN is further increased, corrosion will affect the atmospheric transfer line and HAGO circuits and the bottoms of both towers.

Predictive models of considerable complexity exist but the reliability of such models is not always guaranteed—remembering that such models may have been derived using data from the corrosive character of model compound which are not always representative of the constituents of the naphthenic acid fraction. Furthermore, because of the questionability of the models, it is necessary to apply a monitoring program using hot corrosion probes and hot ultrasonic testing and radiographic testing to confirm the predictions of any active model and to adjust the predicted limits as necessary. While this may seem an unnecessary action, it may permit the use of a simple model which is adjusted on the basis of the real data.

While high TAN crude oils primarily affect the high temperature sections of the atmospheric distillation unit and the vacuum distillation unit, they also affect downstream units as well (such as coker preheaters, catalytic cracker preheater, and hydrotreater preheaters). Some crude oils have also caused overhead system corrosion because of their inefficient performance of the desalting operation.

CHAPTER *4*

Effects in Refining

4.1 INTRODUCTION

The growing variety of discounted opportunity crudes on the market usually contains one or more risks for the purchaser, such as high naphthenic acid content (Chapter 1). As the availability and volume of high acid crude oils processed increase, the risk of experiencing high-temperature corrosion on refinery equipment must be considered. Detailed studies carried out during laboratory and field evaluations, utilizing online monitoring systems, identified associated problems while processing naphthenic acid crudes.

Thus, naphthenic acid corrosion (NAC) can commonly occur in refinery streams operating between 220°C and 400°C (430−750°F) and most typically affects crude and vacuum units (Tebbal, 1999). The corrosion rate of carbon and low-alloy steels is a function not only of the total acid number (TAN) but also of sulfur content, temperature, and fluid flow conditions as well as other factors (Chapter 3). Addition of molybdenum to stainless steel produce alloys with useful resistance to naphthenic acid constituents.

Corrosion at temperatures below 220°C (430°F) has been reported in several circumstances: (1) in atmospheric overhead systems, lighter organic acids—such as acetic acid and formic acid—present in the acidic crudes can cause corrosion, (2) in vacuum overhead systems, light organic acids formed by the degradation of naphthenic acids in the vacuum feed furnace may be present and have also caused corrosion, and (3) in vacuum systems (Figure 4.1). Corrosion at temperatures as low as 180°C has been reported due to true naphthenic acids (Kapusta et al., 2004; Groysman et al., 2005, 2007) although the true boiling point of these acids would be expected to be much higher.

Besides sulfur, crude contains many species that are quantified by the TAN of the oil. This number is not specific to a particular acid but refers to all possible acidic components in the crude and is defined by the amount of potassium hydroxide required to neutralize the acids in

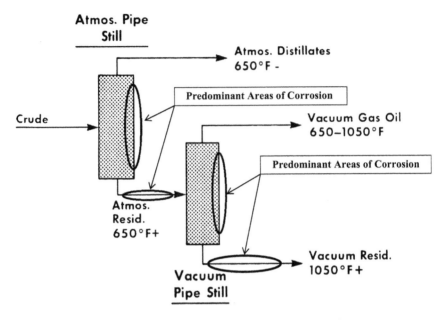

Figure 4.1 Atmospheric and vacuum distillation units showing the predominant areas of NAC.

1 g of oil. Typically found are naphthenic acids, which are organic, but also mineral acids (such as hydrogen sulfide, hydrogen cyanide, and aqueous carbon dioxide) can be present, all of which can contribute significantly to corrosion of equipment. Even materials suitable for sour service do not escape damage under such an onslaught of aggressive compounds. Again, because of cost considerations, a trend toward a preference for crudes with a higher TAN is noticeable.

The variety of high-acid crude oils on the market usually introduces one or more risks for the purchaser. As the availability and volume of highly naphthenic crudes processed increase, the risk of experiencing high-temperature corrosion on refinery equipment must be considered.

In addition to high-temperature corrosion management, many of these high-acid crude oils can be harder to desalt and lead to increased overhead corrosion, fouling, and product stability issues (Kapusta et al., 2004; Groysman et al., 2005, 2007). To be sure, this is another compelling reason to develop and outline the proper evaluation techniques and safe management of naphthenic acid crudes.

Over the next 5 years (2015–2020), it is forecast that high-acid crude oils (crudes having a TAN in excess of >1.0 mg KOH per gram of crude oil) will continue to increase significantly, with production rising across the world. All of these crude oils have significant acid numbers. Therefore, corrosion management is of vital importance to ensure that corrosion risk to 'the plant is minimized, a proper inspection system is in place to identify the corrosion which might occur, and areas of the plant that might be subject to severe corrosion are identified so that the need for more corrosion-resistant alloys can be predicted (Johnson et al., 2003).

4.2 PROCESS EFFECTS

In order to gain an economic advantage, many refiners are looking increasingly at processing high levels of naphthenic crude oils in their crude slates (Johnson et al., 2003). Many high-acid crude oils can cause corrosion in high-temperature regions within the refinery, normally around the crude and vacuum towers.

Corrosion by naphthenic acid constituents is manifested in the form of isolated, deep pits in partially filmed areas and/or impingement attack in essentially film free. Damage is in the form of unexpected high corrosion rates on alloys that would normally be expected to resist sulfidic corrosion. In many cases, even very highly alloyed materials (i.e., 12 Cr, AISI 316 and 317) have been found to exhibit sensitivity to corrosion under these conditions. NAC is differentiated from sulfidic corrosion by the nature of the corrosion (pitting and impingement) and by its severe attack at high velocities in crude distillation units. Crude feedstock heaters, furnaces, transfer lines, feed and reflux sections of columns, atmospheric and vacuum columns, heat exchangers, and condensers are among the types of equipment subject to this type of corrosion.

There are several important variables to consider while performing a risk assessment on a unit such as: stream analysis, temperature, velocity, metallurgy, and flow regimes (Wu et al., 2004a,b). Every piece of the puzzle must be analyzed before the best mitigation strategies can be developed, including: (1) stream analysis, (2) velocity, (3) two-phase flow, (4) areas of turbulence, (5) predictable zones of first vaporization or condensation, (6) reactive sulfur content of the various side-cut oils, (7) metallurgy, (8) other overhead corrosion, desalting and fouling variables, and (9) side-cut stability.

In the stream analysis, the acid number and naphthenic acid content (from the *naphthenic acid titration test* (NAT)) (Chapters 1 and 2) for most crude oils varies with the temperature of distillation fraction. The results for the NAT represent only the naphthenic acids within the TAN. There are many different naphthenic acid species, and some are more corrosive than others. Testing the whole crude and the side cuts shows where the different naphthenic acids will distill and concentrate. It is recommended that such testing is conducted on the anticipated blends that could be encountered to ensure the contributions of other crudes to TAN testing and naphthenic acid titration data are obtained.

Typical desalter problems include effluent water with high oil and solids content (and subsequent problems at the wastewater treating plant), poor desalting efficiency, uncontrollable emulsions, and basic sediment and water (BS&W) carryover into desalted crude. Effective wastewater treatment is another key factor in solving the complex treatment challenge.

The same feedstocks that cause desalter problems can also cause crude unit distillation column overhead corrosion problems due to the higher chloride loadings caused by lower desalter performance. Thermally produced bitumen from the Athabasca oil sands deposits may also have a high TAN, which can cause upgrader or refinery high-temperature NAC problems in the crude unit atmospheric and vacuum distillation systems (Stark et al., 2002; Turini et al., 2011). They can also contribute to crude unit distillation tower overhead corrosion problems, as high TANs promote salt hydrolysis to hydrogen chloride and can thermally degrade to form lower molecular-weight organic acids. These acids can increase both unit neutralizer demand and overhead system corrosion potential.

In addition, to the obvious process such as desalting and distillation, naphthenic acids are implicated in problems with downstream units like coking units, hydrotreaters, and fluid catalytic cracking units (FCCUs). Also, difficulties arise with desalting, and there are risks of increased fouling due to low API gravities of High acid crudes and their contents of asphaltene constituents, naphthenic acids, and calcium naphthenates.

The following sections present descriptions of the various processes that are likely to come into contact with high-acid crudes as well as heavy crude with a tendency to contain high amounts of naphthenic acids (Hau and Mirabal, 1996; Hopkinson and Penuela, 1997).

4.3 DESALTING

Desalting crude oil is the first step in refining that has a direct effect on corrosion and fouling. By mixing and washing the crude with water, salts and solids transfer to the water phase which settles out in a tank. An electrostatic field is induced to speed up the separation of oil and water. In this way, inorganic salts that could cause fouling or hydrolyze and form corrosive acids are largely removed. Often, chemicals are added in the form of demulsifiers to break the oil/water emulsion. Also, chemicals such as caustic soda are introduced to neutralize acidic components. Uncontrolled feeding of caustic can, however, have a detrimental effect. An excess of caustic can result in the formation of soap due to, for instance, the presence of fatty acids. Soap stabilizes the oil water mixture and obstructs the separation process. Also, too strong a mixing of crude and water can create an emulsion that is very difficult to break. Frequently, the crude arrives at the refinery as an emulsion due to the presence of water that had been used to maximize the oil extraction from the oil reservoir, or water might have occurred naturally in the reservoir. It can happen that emulsions are too strong and prove impossible to break. When this is the case, a lot of the contaminants end up in downstream processes, which may have serious consequences.

As a first step in the refining process, crude oil often contains hydrocarbon gases, hydrogen sulfide, carbon dioxide, formation water, inorganic salts, suspended solids, and water-soluble trace metals which must be removed by desalting (dehydration) (Speight and Ozum, 2002; Hsu and Robinson, 2006; Gary et al., 2007; Speight, 2014a). Separators are used to degas produced crude and remove the bulk of formation water. To meet the water content specified by pipeline companies, dehydrators are used to remove much of the remaining formation water and a portion of emulsified water.

The typical methods of crude oil desalting (chemical separation and electrostatic separation) use hot water as the extraction agent. In chemical desalting, water and chemical surfactant (demulsifiers) are added to the crude, heated so that salts and other impurities dissolve into the water or attach to the water, and then held in a tank where they settle out. Electrical desalting is the application of high-voltage electrostatic charges to concentrate suspended water globules in the bottom of the settling tank. Surfactants are added only when the crude has a large amount of suspended solids. Both methods of desalting are continuous. A third and less-common process involves filtering heated crude using diatomaceous earth.

In the process, the feedstock crude oil is heated to between 65°C and 175°C (150°F and 350°F) to reduce viscosity and surface tension for easier mixing and separation of the water—this is particularly important for treating heavy oil. The temperature is limited by the vapor pressure of the crude oil feedstock. In both methods, other chemicals may be added. Ammonia is often used to reduce corrosion. Caustic or acid may be added to adjust the pH of the water wash. Wastewater and contaminants are discharged from the bottom of the settling tank to the wastewater treatment facility. The desalted crude is continuously drawn from the top of the settling tanks and sent to the crude distillation (fractionating) tower.

Desalting chemicals improve overall desalting efficiency, reduce water and solids carryover with desalted crude, and reduce oil carry-under with brine effluent. Most desalting chemicals are demulsifiers that help break up the tight emulsion formed by the mix valve and produce relatively clean phases of desalted crude and brine effluent. Demulsifiers are usually purchased from the same process additives suppliers that supply antifoulants, filming amine corrosion inhibitors, liquid organic neutralizers and similar products for controlling overhead corrosion, and fouling problems on crude units (or elsewhere in the refinery). If necessary, demulsifiers can be custom formulated for high water removal rates from crudes, but at the cost of poor solids wetting and oil carry-under with the brine discharge. They can also be formulated for high oil removal rates from brine, but at the cost of water carryover with desalted crude. When formulated for high solids wetting rates, brine quality often decreases and water carryover with desalted crude increases.

Chemicals such as caustic soda are also introduced to neutralize acidic components, which is not always successful in terms of naphthenic acid removal. Uncontrolled feeding of caustic can, however, have a detrimental effect. An excess of caustic can result in the formation of soap due to, for instance, the presence of fatty acids. Soap stabilizes the oil water mixture and obstructs the separation process. Also, too strong a mixing of crude and water can create an emulsion that is very difficult to break. Frequently, the crude arrives at the refinery as an emulsion due to the presence of water that had been used to maximize the oil extraction from the oil reservoir, or water might have occurred naturally in the reservoir. It can happen that emulsions are too strong and prove impossible to break and the contaminants end up in downstream processes, which may have serious consequences.

One process parameter that can play a vital role in both neutralizing acids and demulsification is process pH. Careful monitoring of the pH in the desalter water effluent allows for efficient dosing of caustic or acid which may result in significant cost savings. The stability of the oil/water emulsion depends partly on pH. Maintaining the pH of the mixture within a certain range helps the demulsifier chemicals in breaking the emulsion by interacting directly with the water droplets. The speed and quality of the separation process can thus be improved which leads to less water carryover, which in turn can result in a significant reduction in downstream acid corrosion.

However, naphthenic acid crude constituents in crude oil have natural emulsification tendencies. As the pH of the water inside the desalter increases, the sodium naphthenates can form very stable emulsions. Maintaining an acidic effluent desalter water is important to combat the role sodium naphthenates play in desalter upsets.

Processing crude oils containing high levels of calcium naphthenates can present a number of operating challenges (Piehl, 1988). Two processing technologies can help refiners successfully process these crudes: (1) use of a metals removal technology developed to remove calcium in the crude unit desalting operation and (2) chemical treatments in the crude and vacuum columns.

Several crude oils have come into production within the last few years that contain high levels of calcium naphthenates. Typically, these crudes are medium to heavy (specific gravity 0.89−0.95, i.e., 17.5−27.5 API), highly biodegraded oils, high in naphthenic acid content, and containing high concentrations of calcium ion in the formation water. Generally, calcium naphthenates found in many crude oils are insoluble in oil, water, and organic solvents, and this can lead to fouling in separators, hydrocyclones, heat exchangers, and other upstream production equipment. When blended into refinery crude oil feedstocks, these crude oils can create a number of processing and product quality challenges in the tank farm, crude unit, and downstream units. These processing issues result from several observed attributes of crude oil blends containing calcium naphthenates: (1) high-conductivity crude blends, (2) tendency to form stable emulsions, (3) high calcium content of atmospheric and vacuum residua, (4) higher levels of low-molecular-weight organic acids in crude unit distillation column overheads, and (5) increased high-temperature NAC activity.

In terms of crude oil conductivity (Speight, 2014a), some high-acid crude oils have a sufficiently high conductivity to interfere with electric desalters to the point where dehydration is inefficient. In addition to the corrosion potential from naphthenic acid constituents in the crude, fouling can also occur in downstream units due to corrosion by-products. The corrosion by-product of NAC is iron naphthenate [$Fe(naphthenate)_2$] (Chapter 1). Corrosion mitigation is required to prevent premature fouling/cleanings due to the buildup of NAC by-products.

Desalting of high-acid heavy crude oils is much more challenging, and many older desalter systems will need to be enlarged or replaced with one utilizing a newer desalting technology. Desalter performance can be hampered by factors, such as the increased salt content in the heavy crudes, as well as the high crude density and viscosity, that make oil–water separation more difficult. Moreover, the combination of asphaltene precipitation and high naphthenic acid concentration will increase the tendency to form stable water-in-oil emulsion (rag layer), and potentially cause high oil entrainment in brine water and difficulty in maintaining the desired BS&W and salt removal at the desalter outlet.

One of the critical aspects of preventing desalter upsets is the ability to detect and monitor the emulsion because of the presence of an undesirable mixture of dispersed oil, water, and solids (*rag layer*). The use of conventional level measurement devices, such as guided wave radar and displacer float column, has not been proven as accurate and reliable in stabilizing the desalter interface levels. The poor rag layer detection would also often result in relatively high chemical injection rates to control the rag layer. Instrumentation for level detection (when three phases exist) is available to accurately monitor the rag layer to optimize the desalter performance.

4.4 DISTILLATION

The first step in the refining process—after the desalting step—is the separation of crude oil into various fractions or straight-run cuts by distillation in atmospheric and vacuum towers. The main fractions or "cuts" obtained have specific boiling point ranges and can be classified in order of decreasing volatility into gases, light distillates, middle distillates, gas oils, and residuum.

As the crude oil slate to refineries include heavier crude oils—typically high-acid crude oils—the atmospheric tower and vacuum tower distillate cut points tend to suffer due to increasing difficulty of vaporization. Therefore, changes can be made such as increasing the temperature and residue stripping efficiency, lowering the pressure and flash zone oil partial pressure, and modifying the pump-around protocols. For the atmospheric unit, other key areas include the oil preheat train and charge furnace, column internals, and metallurgy of the unit exposed to higher sulfur and naphthenic acids. For the vacuum unit, evaluations should be made of the furnace sizing and outlet temperature, decoking capability, wash-zone capacity, and steam requirement (if it is a wet vacuum column). Deep-cut vacuum distillation via a revamp of the unit to cut deeper into the resid to make additional feedstock for the FCCU and/or for hydrocracking unit is always an attractive option to produce higher yield of liquid fuel precursors.

However, such changes can not only lead to NAC but also to an increase in the rate of corrosion (Mottram and Hathaway, 1971; Blanco and Hopkinson, 1983).

When sulfur is present, iron sulfide scales are formed by sulfur corrosion on the inner walls of refinery distilling towers and transfer lines operating on sour crude oils. Sulfide scales are generally considered to partially reduce corrosion by other corrosive species in crudes, especially naphthenic acids. Sulfur and NAC occur simultaneously at similar high temperatures in both atmospheric and vacuum distillation unit.

Iron sulfide scale typically forms a semiprotective barrier against naphthenic acid attack. The scale can be removed by high wall shear stress (e.g., high velocity), which exposes the fresh metal beneath to further corrosion. Naphthenic acids can also convert iron sulfide to oil-soluble iron naphthenate, which weakens and helps remove the scale. The presence of more active sulfur species (such as hydrogen sulfide) tends to stabilize the sulfide scale against this latter form of attack. The net result of these effects is that NAC behavior can be time variant, localized, and difficult to predict.

4.4.1 Atmospheric Distillation

At the refinery, the desalted crude feedstock is preheated using recovered process heat. The feedstock then flows to a direct-fired crude charge heater where it is fed into the vertical distillation column just above the bottom, at pressures slightly above atmospheric and at temperatures ranging from 345°C to 370°C (650−700°F)—heating crude oil above these temperatures may cause undesirable thermal cracking. All but the highest boiling fractions flash into vapor. As the hot vapor rises in the tower, its temperature is reduced. Heavy fuel oil or asphalt residue is taken from the bottom. At successively higher points on the tower, the various major products including lubricating oil, heating oil, kerosene, gasoline, and uncondensed gases (which condense at lower temperatures) are drawn off.

The fractionating tower, a steel cylinder about 120 ft high, contains horizontal steel trays for separating and collecting the liquids. At each tray, vapors from below enter perforations and bubble caps (Speight, 2014a). They permit the vapors to bubble through the liquid on the tray, causing some condensation at the temperature of that tray. An overflow pipe drains the condensed liquids from each tray back to the tray below, where the higher temperature causes reevaporation. The evaporation, condensing, and scrubbing operation is repeated many times until the desired degree of product purity is reached. Then side streams from certain trays are taken off to obtain the desired fractions. Products ranging from uncondensed fixed gases at the top to heavy fuel oils at the bottom can be taken continuously from a fractionating tower. Steam is often used in towers to lower the vapor pressure and create a partial vacuum. The distillation process separates the major constituents of crude oil into so-called straight-run products. Sometimes crude oil is *topped* by distilling off only the lighter fractions, leaving a heavy residue that is often distilled further under high vacuum.

Many areas of the crude distillation unit can be susceptible to high-temperature NAC. These areas can most simply be identified as those which: (1) are exposed to hydrocarbon fluids that contain corrosive levels of naphthenic acids, (2) operate at temperatures in the range 220–400°C (425–750°F), and (3) are constructed with metallurgy not generally considered to be resistant to NAC attack. Stainless steels such as 316, 316L, 317, or 317L are generally considered to be resistant materials. Additionally, areas of the atmospheric distillation unit that are susceptible to NAC according to the above parameters typically include: (1) hot feedstock preheat exchanger network, (2) atmospheric tower heater tubes, (3) atmospheric tower transfer lines, (4) the lower section of the atmospheric tower including lining, trays and associated atmospheric gas oil pump around/product draw system, and atmospheric tower bottoms line and any bottoms heat exchangers, if not integrated with vacuum unit.

4.4.2 Vacuum Distillation

In order to further distill the residuum or topped crude from the atmospheric tower at higher temperatures, reduced pressure is required to prevent thermal cracking. The process takes place in one or more vacuum distillation towers. The principles of vacuum distillation resemble those of fractional distillation and, except that larger diameter columns are used to maintain comparable vapor velocities at the reduced pressures, the equipment is also similar. The internal designs of some vacuum towers are different from atmospheric towers in that random packing and demister pads are used instead of trays. A typical first-phase vacuum tower may produce gas oils, lubricating oil base stocks, and heavy residual for propane deasphalting. A second-phase tower operating at lower vacuum may distill surplus residuum from the atmospheric tower, which is not used for lube-stock processing, and surplus residuum from the first vacuum tower not used for deasphalting. Vacuum towers are typically used to separate catalytic cracking feedstock from surplus residuum.

4.4.3 Other Areas

Within refineries, there are numerous other, smaller distillation units designed to separate specific and unique products. Columns all work on the same principles as the towers described above. For example, a depropanizer is a small column designed to separate propane and lighter gases from butane and heavier components. Another larger column is used to separate ethyl benzene and xylene. Small "bubble"

towers called strippers use steam to remove trace amounts of light products from heavier product streams.

In *furnace tubes and transfer lines*, vaporization and fluid velocity
are very high. The high-temperature conditions appear to activate even
small amounts of naphthenic acid in oil increasing corrosion significantly. Thus, at furnace tubes and transfer lines conditions, the influence of temperature, velocity, and degree of vaporization is very large.
Process conditions such as load and steam rate and especially turbulence affect corrosivity (Craig, 1995, 1996).

The ancillary areas of the vacuum distillation unit that are susceptible to NAC according to the above parameters typically include: (1)
the vacuum heater tubes, (2) the vacuum tower transfer lines, (3) the
vacuum tower itself—the lining, trays, packing—and associated light
vacuum gas oil and heavy vacuum gas oil pump around/product draw
systems, (4) the vacuum tower over flash draw and pump-back lines
and associated equipment, and (5) the vacuum tower bottoms line and
heat exchangers. Other areas of the unit may also be susceptible
depending on crude blend.

4.4.4 Effects of Naphthenic Acids

High-temperature corrosivity of distillation units due to the presence of
naphthenic acids is a major concern to the refining industry. The difference in process conditions, materials of construction, and blend processed
in each refinery and especially the frequent variation in the crude slate
increases the problem of correlating corrosion of a unit to a certain type
of crude oil. In addition, a large number of interdependent parameters
influence the high-temperature crude corrosion process (Chapter 3).

Briefly, processing high-acid crude oils will increase the potential
for NAC in crude oil distillation units. If not controlled, high-
temperature NAC will result in higher equipment replacement costs,
lower unit reliability and availability, and increased severity of downstream unit fouling due to elevated levels of iron naphthenates in crude
unit distillates—which can also affect color stability in distillates from
the atmospheric distillation unit.

Moreover, despite a good desalting operation, corrosive substances
such as naphthenic acids (which are not typically removed by a caustic
wash) can still appear during downstream processing. As an example, the

sour (acidic) water corrosion that occurs in the atmospheric tower if produced from the steam, which is injected into the tower to improve the fractionation, condenses in the upper part of the unit. Acidic substances will dissolve in the condensate to form an acidic liquid which will cause corrosion in the top section of the tower and in the overhead condenser. This may lead to a requirement of frequent retubing of the condenser and in severe cases to replacement of the entire crude tower top.

For the atmospheric unit, other key areas include the oil preheat train and charge furnace, column internals, and the metallurgy of the unit exposed to higher sulfur and high TAN. For the vacuum towers, evaluations should be made regarding furnace sizing and outlet temperature, decoking capability, wash-zone capacity, and steam requirement (if it is a wet column). Deep-cut vacuum distillation via a revamp of the unit to cut deeper into the resid, to make additional feedstock for a fluid catalytic cracking unit or for a hydrocracking unit feed, is one of the first and most attractive options a refiner should consider.

NAC occurs primarily in high-velocity areas of crude distillation units in the 220–400°C (430–750°F) temperature range. Lesser amounts of corrosion damage are found at temperatures greater than 400°C (750°F), probably due to the decomposition of naphthenic acids or protection from the coke formed at the metal surface. Velocity and, more important, wall shear stress are the main parameters affecting NAC. Fluid flow velocity lacks predictive capabilities. Data related to fluid flow parameters, such as wall shear stress and the Reynold's Number, are more accurate because the density and viscosity of liquid and vapor in the pipe, the degree of vaporization in the pipe, and the pipe diameter are also taken into account. Corrosion rates are directly proportional to shear stress. Typically, the higher the acid content the greater the sensitivity to velocity. When combined with high temperature and high velocity, even very low levels of naphthenic acid may result in very high corrosion rates.

Most importantly, NAC activity is dependent upon a number of key variables, which include but (depending upon the crude oil slate and the refinery equipment) are not limited to: (1) the naphthenic acid content of the feedstocks—acid-based corrosion is either reduced or augmented depending on high or low TAN, (2) sulfur content, (3) sulfur types, (4) feedstock phase—fluid or vapor, (5) the temperature of the metal surfaces being contacted by the corrosive feedstock constituents, (6) decomposition of the naphthenic acids to lower molecular weight acids, (7) condensation of the authentic acids from the vapor phase into a liquid phase, such as condensation and dissolution, and (8) the shear stress of the hydrocarbon moving across the metal surface, which is a function of velocity and turbulence of the flowing stream.

In terms of the fluid velocity, at low velocity, acid concentration caused by boiling and condensing causes corrosive attack, whereas at high velocity multiphase stream rapid corrosion can occur due to erosion−corrosion. Furthermore, NAC is accelerated in furnaces and transfer lines where the velocity of the liquid/vapor phase is increased. Areas subject to high turbulence of the fluids are also subject to severe corrosion. In addition, turbulence and cavitation in pumps may result in rapid attack, and the type of alloy in use for high-acid crude oils as well as surface temperature and shear stresses can also render the system susceptible to corrosion by naphthenic acid attack—in some situations 316 stainless steel, 317 stainless steel, and high-molybdenum alloys (more molybdenum) may offer more resistance to NAC. But it must be remembered that the effect of the whole system—such as the types of crude oil, the chemical structure of the naphthenic acids, and the equipment—all play a role in determining the occurrence and extent of the corrosion (Chapters 1 and 2) (Speight, 2014b).

NAC is differentiated from sulfidic corrosion by the nature of the corrosion (pitting and impingement) and by its severe attack at high velocities in crude distillation units. Crude feedstock heaters, furnaces, transfer lines, feed and reflux sections of columns, atmospheric and vacuum columns, heat exchangers, and condensers are among the types of equipment subject to this type of corrosion. However, at high temperatures, especially in furnaces and transfer lines, the presence of naphthenic acids may increase the severity of sulfidic corrosion. The presence of these organic acids may disrupt the sulfide film, thereby promoting sulfidic corrosion on alloys that would normally be expected to resist this form of attack (i.e., 12 Cr and higher alloys). In some cases, such as in side-cut piping, the sulfide film produced by hydrogen sulfide is believed to offer some degree of protection from NAC. In general, the corrosion rate of all alloys in the distillation units increases with an increase in temperature.

The presence of sulfur compounds with the naphthenic acids considerably increases corrosion in the high-temperature parts of the distillation units (Chapters 1 and 2). Isolated deep pits in partially passivated areas and/or impingement attack in essentially passivation-free areas are typical of NAC. Damage is in the form of unexpected high corrosion rates on alloys that would normally be expected to resist sulfidic corrosion. In many cases, even very highly alloyed materials (i.e., 12 Cr, AISI types 316 and 317) have been found to exhibit sensitivity to corrosion under these conditions.

The top section of a crude unit can be subjected to a multitude of corrosive species. Hydrochloric acid, formed from the hydrolysis of calcium and magnesium chlorides, is the principal strong acid responsible for corrosion in the crude unit top section. Carbon dioxide is released from crudes typically produced in petroleum recovery operations that involve the use of carbon dioxide flooding fields as well as crude oils that contain a high content of naphthenic acid.

Mitigation of NAC through process changes includes any action to remove or at least reduce the amount of acid (and acid gases) gas present and to prevent the accumulation of water on the tower trays. Material upgrading includes lining of distillation tower tops with alloys resistant to hydrochloric acid. Design changes are used to prevent the accumulation of water—these include redesign of the coalescers and water draws. The application of chemicals includes the injection of a neutralizer to increase the pH and a corrosion inhibitor. The presence of many weak acids, such as fatty acids and carbon dioxide, can buffer the environment and require greater use of neutralizers. Excess neutralizers may cause plugging of trays and corrosion under the salt deposits.

A dew point probe is typically placed in a location at least 38°C (100°F) above the calculated dew point temperature. The probe elements are then cooled internally by cold air injection and the temperature at which the first liquid drop forms is determined for the actual conditions in the tower. The injection point and the amount of chemicals used depend on the knowledge of the temperature in the tower where condensation starts. With the number of corrosive species present, the calculated dew point may be much lower than the actual dew point.

Processing crude oils that have a high content of calcium naphthenate derivatives can, as with many high-acid crude oils, result in higher loadings of low-molecular-weight organic acids and carbon dioxide in the upper portions of the crude and vacuum columns and overhead condensing systems. The amount and distribution of lower molecular weight acids and carbon dioxide in these systems is a function of the distribution of organic acid molecular weights in the crude oil, plus heater outlet, side cut, and column overhead temperatures.

In addition to naphthenic acids, the overhead of an atmospheric distillation tower crude unit can be subjected to a multitude of corrosive species: (1) hydrochloric acid, formed from the hydrolysis of calcium and magnesium chlorides, is the principal strong acid responsible for corrosion in crude unit overhead, (2) carbon dioxide, released from crudes typically produced in enhanced oil recovery system using carbon dioxide as the recovery gas flooded fields, (3) hydrogen sulfide, released from sour crudes, increases significantly corrosion of crude unit overhead, (4) sulfuric acid and sulfurous acid, formed by either oxidation of hydrogen sulfide or direct condensation of sulfur dioxide and sulfur trioxide, and (5) low-molecular fatty acids such as formic acid, HCO_2H, acetic acid, CH_3CO_2H, propionic acid, $CH_3CH_2CO_2H$, and butanoic acid, $CH_3CH_2CH_2CO_2H$, which are released from crude oils with a high content of naphthenic acid. Any of these acids coming into contact with water in condensation areas will be increased in the corrosivity potential. Furthermore, the presence in the feedstock of the lower molecular acids can buffer the environment and require a higher use of neutralizing chemicals. However, excessive amounts of a neutralizer chemical may cause plugging of trays and corrosion under the salt deposits.

Mitigation of this type of corrosion is performed by process changes, material upgrading, design changes, and injection of chemicals such as neutralizers and corrosion inhibitors (Petkova et al., 2009). Process changes include any action to remove or at least reduce the amount of acid gas present and to prevent accumulation of water on the tower trays. Material upgrading includes lining of distillation tower tops with alloys resistant to hydrochloric acid. Design changes are used to prevent the accumulation of water and include coalescers and water draws. The application of chemicals includes the injection of a neutralizer to increase the pH and a corrosion inhibitor. The presence of many weak acids such as fatty acids and carbon dioxide can buffer the environment and require a higher use of neutralizers.

Typically, corrosion inhibitors and neutralizers such as caustic soda or ammonia are injected with the aim of increasing the pH of the sour water.

Although this is an obvious response to the problem, it is not always advisable—excess of neutralizer may cause plugging of trays and corrosion under the salt deposits. For example, the presence of various acid gases and ammonia can result in salt deposition − ammonium bisulfide (NH_4HS) is one of the main causes of sour water corrosion. Alkalinity of the water (pH >7.6) dramatically increases ammonium bisulfide corrosion. Overdosing the amount of caustic is easily achieved—as in the desalting operation, the key to corrosion reduction is in accurate pH control. The application of dew point equipment may offer some benefits for mitigating corrosion. The dew point probe is typically placed in a location at least 38°C (100°F) above the calculated dew point temperature. The probe elements are then cooled internally by cold air injection and the temperature at which the first liquid drop forms is determined for the actual conditions in the tower. The injection point and the amount of chemicals used depend on the knowledge of the temperature in the tower where condensation occurs. With the number of corrosive species present, the calculated dew point may be much lower than the actual dew point.

An additional concern for chemical treatment in the atmospheric distillation unit overhead is the application of the *film technology* in which the corrosion inhibitor forms a thin film on the metallurgy and prevents corrosion. However, if the film-forming inhibitor has surface properties this can cause a water emulsion to occur in the overhead stream (typically a naphtha stream). The water in the stream can cause further corrosion problems downstream of the distillation unit—careful selection of corrosion inhibitors to minimize this effect should be exercised.

Metallurgy will have an impact on the atmospheric and vacuum units. There are two major causes for concern: sulfidic attack due to increased sulfur content and (2) NAC, since most heavy crudes result in sulfidation of the metal as well as naphthenic acid attack. The most common solution to the NAC problem is increased metallurgy in the affected equipment to 317L stainless steel or alloys with approximately 3% w/w molybdenum.

One key parameter that can be very expensive is the transfer temperature for products to the downstream units. If the transfer piping is carbon steel, it is important to maintain the unit transfer temperature below the temperature at which NAC is a concern, typically approximately 205−235°C (400−450°F).

Furthermore, along with the high propensity for cracking, processing high-acid heavy sour crudes can lead to a high rate of coke formation, typically concentrated in the vacuum tower heater furnace. To minimize coke formation, several key mitigation strategies should be incorporated into the design such as the use of high fluid velocity in the heater tubes. In addition, a double-fired vacuum heater design will reduce the peak heat flux in the tubes to minimize the coking potential. Also, stripping steam in the vacuum column shows that this will minimize the required vacuum heater outlet temperature for a fixed vacuum resid cut point and vacuum column diameter.

Several design issues can affect the design and operation of the crude preheat exchanger train. High-acid heavy crude oils, which typically have a high viscosity that significantly impairs the heat transfer in the cold preheat exchangers, and options to improve heat transfer with varying baffle configurations or twisted tube bundles are available.

4.5 VISBREAKING

The visbreaking process is a thermal (noncatalytic) process that was originally developed to reduce the resid viscosity to meet the specifications for heavy fuel oil. In the process, the high-boiling feedstock (residuum, heavy oil, tar sand bitumen) is converted to distillable products. The thermal reactions are not allowed to proceed to completion, and the hot reaction mix is quenched with a lower boiling (gas oil type) product (Speight, 2014a). In modern refineries, the process is used to frequently to convert heavy feedstocks into fuel oil into valuable gasoline and gas oil and produces residual fuel oil sold as marine fuel (Speight and Ozum, 2002; Hsu and Robinson, 2006; Gary et al., 2007; Speight, 2014a). The conversion is low because the process takes place just before the point of coke formation.

In the process, the feedstock is heated 430−510°C (800−950°F) at atmospheric pressure and mildly cracked in a heater. It is then quenched with cool gas oil to control excessive cracking and flashed in a distillation tower. The thermally cracked residual material, which accumulates in the bottom of the fractionation tower, is vacuum flashed in a stripper and the distillate recycled.

The introduction of opportunity crudes, including high-acid crudes, to refinery slates poses additional challenges for visbreaker optimization.

Experience in dealing with frequently changing complex blends has identified unappreciated problematic feed characteristics that can severely limit visbreaker performance due to incompatibility phenomena or (more pertinent to the present text) corrosion (Rijkaart et al., 2009; Speight, 2014a).

A major concern with processing high-acid crude oils is blending and mixtures, since many of the high-acid (heavy) crude oils may be incompatible with other crudes schedule for blending and use by the refinery (Speight, 2014a). Therefore, it is important to use the oil compatibility test methods to predict the proportions and order of blending of oils that would prevent incompatibility prior to the purchase and scheduling of crudes (Speight, 2014a).

High-acid crude oils may compete with heavy, sour crudes requiring more residual processing capacity, such as visbreaking as a pretreatment process prior to sending the visbroken feedstock for further processing. On the other hand, some of the tar sand bitumen blends have high-acid content that requires investment in improved metallurgy and chemical additives. No matter which resid conversion technology is selected, coke and sediment formation as well as NAC are often the major concerns.

Organic acids formed by decomposition of naphthenic acids and phenol derivatives (Chapter 2) can cause significant metal loss in visbreaker units (O'Kane et al., 2010a). Corrosion inside the units can part way up the visbreaker fractionator column on the trays and within the downcomers, and the corrosion is manifested by the presence of smooth, uniform zones of metal loss characteristic of organic acid corrosion. Hydrogen flux measurement can be used to track the activity of NAC at 200°C (390°F)—in the visbreaker fractionator, as in other locations where naphthenic acids occur, the only source of hydrogen in the process stream is likely to be from the cracking of hydrocarbons (Zetlmeisl, 1996; Dean and Powell, 2006; O'Kane et al., 2010a,b; Rudd et al., 2010).

In some visbreakers, smooth, uniform corrosion in highly localized areas has been found within the downcomer trays, with the amount of the corrosion varying depending on the position of the trays. The corrosion is typically localized to areas of liquid flow, and the pattern of the corrosion suggests a relation to the fluid flow patterns (O'Kane et al., 2010a,b). It has been reported that the operating temperatures at the corroded trays were within the lower end of the range at which

NAC has been observed. Additionally, the corrosivity of naphthenic acids was known to be strongly velocity related (Babaian-Kibala et al., 1993; Nugent and Dobis, 1998; Babaian-Kibala and Nugent, 1999). When this occurs, the visual appearance of the corroded areas is generally consistent with corrosion by organic acids—specifically, the affected area may have no adherent scale also there may be bright patches that appear freshly corroded along with a pattern of surface roughness that suggests the corrosion is flow related.

However, any naphthenic acids in the feed to a visbreaker unit would need to be degraded (cracked) to produce lighter acids in order to reach this upper portion of the column. It is possible for several lower molecular weight organic acids to be formed from naphthenic acids at visbreaking temperatures and the boiling points of the individual lower molecular weight acids may correspond well with the temperature in the area of highest corrosion, thereby indicating the curative agent for the corrosion. Other corrosion sites within the unit may also indicate the severity or lack of severity of the corrosion that is due to acid species that are present in lesser amount than the main corrosive agent.

4.6 COKING

Coking is a severe method of thermal cracking used to convert high-boiling residua as well as heavy oil and tar sand bitumen into lower boiling products or distillates. Unlike visbreaking in which the thermal reactions are not allowed to proceed to completion, coking is a severe method of thermal cracking in which cracking to extinction is practised. Coking produces straight-run gasoline (coker naphtha) and various middle-distillate fractions used as catalytic cracking feedstock as well as coke. The two most common processes are delayed coking and continuous (contact or fluid) coking. Three typical types of coke are obtained (sponge coke, honeycomb coke, and needle coke) depending upon the reaction mechanism, time, temperature, and the crude feedstock.

4.6.1 Delayed Coking

Delayed coking, a carbon rejection technology, is the most popular way to upgrade heavy feedstocks. However, in the process, approximately 70–80% v/v of the total heavy feedstocks is converted into valuable transportation fuels while the remaining portion is downgraded to coke. There are three limitations to achieving higher liquid yields: (1) secondary cracking of valuable volatile liquid products, (2) a combination of smaller-ring aromatics to form polynuclear aromatics (PNAs), and (3) the formation of PNAs via aromatization of hydroaromatic constituents. To minimize the secondary cracking of volatile liquid products, a coker reactor with a very short vapor residence time but a lengthy resid residence time is preferred to achieve complete conversion to coke. The latest developments in the conventional delayed coking technology emphasize design and operation of major equipment (e.g., coke drums, heaters, and fractionators), coke quality and yield flexibility, and operability and safety (Radovanović and Speight, 2011).

In the process, the heated charge (typically residuum from atmospheric and vacuum distillation towers) is transferred to large coke drums which provide the long residence time needed to allow the cracking reactions to proceed to completion. Initially, the heavy feedstock is fed to a furnace which heats the residuum to high temperatures (900−950°F) at low pressures (25−30 psi) and is designed and controlled to prevent premature coking in the heater tubes. The mixture is passed from the heater to one or more coker drums where the hot material is held approximately 24 h (delayed) at pressures of 25−75 psi, until it cracks into lighter products. Vapors from the drums are returned to a fractionator where gas, naphtha, and gas oils are separated out. The heavier hydrocarbons produced in the fractionator are recycled through the furnace.

After the coke reaches a predetermined level in one drum, the flow is diverted to another drum to maintain continuous operation. The full drum is steamed to strip out uncracked hydrocarbons, cooled by water injection, and decoked by mechanical or hydraulic methods. The coke is mechanically removed by an auger rising from the bottom of the drum. Hydraulic decoking consists of fracturing the coke bed with high-pressure water ejected from a rotating cutter.

The potential exists for exposure to hazardous gases such as hydrogen sulfide and carbon monoxide, and trace PNAs associated with coking operations. When coke is moved as a slurry, oxygen depletion may occur within confined spaces such as storage silos, since wet carbon will adsorb oxygen. Wastewater may be highly alkaline and contain oil, sulfides, ammonia, and/or phenol. The potential exists in the coking process for exposure to burns when handling hot coke or in the event of a steam-line leak, or from steam, hot water, hot coke, or hot slurry that may be expelled when opening cokers. Safe work practices and/or the use of appropriate personal protective equipment may be needed for exposures to chemicals and other hazards such as heat and noise, and during process sampling, inspection, maintenance, and turn-around activities.

When sour and/or high-acid crude oils—including resids where distillation has concentrated naphthenic acids into the resids (Speight and Francisco, 1990)—are processed in a delayed coking unit, corrosion can occur where metal temperatures are between 265°C and 485°C (450°F and 900°F). At temperatures in excess of 485°C (900°F), coke forms a protective layer on the unit/reactor metal. Nevertheless, the furnace, soaking drums, lower part of the tower, and high-temperature exchangers are usually subject to corrosion. Hydrogen sulfide corrosion in coking can also occur when temperatures are not properly controlled above 485°C (900°F). NAC in delayed coking units has also been reported in the light cycle gas oils and heavy cycle gas oil streams on delayed coking units processing high-acid feedstocks.

4.6.2 Fluid Coking

In the fluid coking process, cracking occurs by using heat transferred from hot, recycled coke particles to feedstock in a radial mixer, called a reactor, at a pressure of 50 psi. Gases and vapors are taken from the reactor, quenched to stop any further reaction, and fractionated. The reacted coke enters a surge drum and is lifted to a feeder and classifier where the larger coke particles are removed as product. The remaining coke is dropped into the preheater for recycling with feedstock. Coking occurs both in the reactor and in the surge drum. The process is automatic in that there is a continuous flow of coke and feedstock.

4.6.3 Effect of Naphthenic Acids

A coking process is one of several options for resid, heavy oil, and tsar sand bitumen processing. The function of the coking unit function is to upgrade the heavy feedstock(s) into more valuable liquid products, and heavy sour crudes have significantly higher amounts of vacuum resid in the feed, the coking unit (especially the delayed coker) is typically one of the most overwhelmed in terms of capacity.

Although debottlenecking is usually possible by improvements to drum cycle time, most refiners will find that the capacity of their existing coking unit is soon overwhelmed by the amount of heavy feed-stock (containing resid naphthenic acids or heavy oil/tar sand bitumen naphthenic acids) that needs to be processed. Therefore, additional coking capacity in the form of a new unit is usually required to significantly increase the amount of high-acid heavy feedstocks processed by the refinery.

Most coking units are constructed of alloy that will resist sulfidic corrosion, but if the feed increases significantly in TAN (naphthenic acid content), some metallurgy reviews are needed. Fortunately, in many cases, the heater temperatures cause the naphthenic acids to decompose, and it may only occur in the feed area after some preheating that is subject to the risk of corrosion.

4.7 CATALYTIC CRACKING

Catalytic cracking decomposes complex high-molecular-weight feed-stocks into less complex low-molecular-weight products in order to increase the quality and quantity of lower molecular weight products and decrease the amount of coke.

Catalytic cracking is similar to thermal cracking except that catalysts facilitate the conversion of the heavier molecules into lighter products. Use of a catalyst (a material that assists a chemical reaction but does not take part in it) in the cracking reaction increases the yield of improved quality products under much less severe operating conditions than in thermal cracking. Typical temperatures are from 455°C to 530°C (850–950°F) at much lower pressures of 10–20 psi. The catalysts used in refinery cracking units are typically solid materials (zeolite, aluminum hydrosilicate, treated bentonite clay, fuller's earth, bauxite, and silica–alumina) that come in the form of powders, beads, pellets, or shaped materials (extrudates).

4.7.1 Fluid Catalytic Cracking

A typical fluid catalytic cracking process involves mixing a preheated hydrocarbon charge with hot, regenerated catalyst as it enters the riser leading to the reactor. The charge is combined with a recycle stream within the riser, vaporized, and raised to reactor temperature (485–540°C; 900–1000°F) by the hot catalyst. As the mixture travels up the riser, the charge is cracked at 10–30 psi. In the more modern

FCCUs, all cracking takes place in the riser. The reactor no longer functions as a reactor but serves as a holding vessel for the cyclones. This cracking continues until the oil vapors are separated from the catalyst in the reactor cyclones. The resultant product stream (cracked product) is then charged to a fractionating column where it is separated into fractions, and some of the heavy oil is recycled to the riser.

Spent catalyst is regenerated to get rid of coke that collects on the catalyst during the process. Spent catalyst flows through the catalyst stripper to the regenerator, where most of the coke deposits burn off at the bottom where preheated air and spent catalyst are mixed. Fresh catalyst is added and worn-out catalyst removed to optimize the cracking process.

NAC takes place where both liquid and vapor phases exist and at areas subject to local cooling such as nozzles and platform supports. Furthermore, when the feedstock is a high-nitrogen feedstock (as is often the case when heavy oil and tars sand bitumen are included as feedstock), exposure to ammonia and cyanide may occur, subjecting carbon steel equipment in the fluid catalytic cracking overhead system to corrosion, cracking, or hydrogen blistering. These effects may be minimized by water wash or corrosion inhibitors. Water wash may also be used to protect overhead condensers in the main column subjected to fouling from ammonium hydrosulfide. Inspections should include checking for leaks due to erosion or other malfunctions such as catalyst buildup on the expanders, coking in the overhead feeder lines from feedstock residues, and other unusual operating conditions.

4.7.2 Effect of Naphthenic Acids

The catalytic cracking process has also become one of several options for resid, heavy oil, and tsar sand bitumen processing. The function of the unit is to upgrade the heavy feedstock(s) into more valuable liquid products, and heavy feedstocks crudes have significantly higher amounts of resid in the feed, the catalytic cracking unit is often also overwhelmed in terms of capacity. In addition, the amount of heavy feedstock (containing resid naphthenic acids or heavy oil/tar sand bitumen naphthenic acids) that needs to be processed continued to increase. Therefore, additional cracking capacity is required to significantly increase the amount of high-acid heavy feedstocks processed by the refinery.

Most cracking units are constructed of alloy that will resist sulfidic corrosion, but if the feed increases significantly in TAN (naphthenic acid content), some metallurgy reviews are required. Fortunately, as with other high-temperature units, the heater temperatures cause the naphthenic acids to decompose, and it may only occur in the feed area after some preheating that is subject to the risk of corrosion. While NAC is normally not a concern much below 200°C (392°F) (Chapters 1−3), as temperature increases in the feedstock pipes to the unit, the corrosion rates may increase until the temperatures are sufficiently high enough (usually on the order of 420°C, 790°F) to decompose the naphthenic acids to lower molecular weight organic acids. The differential between the TAN and the naphthenic acid titration number (Chapter 1) begins to widen with the naphthenic acid titration number decreasing.

However, as with all units, the major concern with the catalytic cracking units and the ability of the units to accommodate high-acid crudes is related to the metallurgy. NAC can be a major concern while the feedstock is hot. However, many refineries transport the feedstocks oils between units at moderate temperatures (approximately 205°C, 400°F) that alleviate the need for upgraded interconnecting pipe between units. In the catalytic cracking unit, the feedstock is usually heated with the catalyst during introduction into the reactor. Once the feedstock is mixed with the catalyst, this might have an adverse effect on stainless steel alloys (even those without significant molybdenum contents) from the acidic attack of the naphthenic acid. Therefore, a careful review of the unit configuration and metallurgy is required to ensure that either upgraded metallurgy is provided or that catalyst mixed with the oil before introduction into the unit does not enhance NAC. It should be recalled that a corrosion by-product of NAC is iron naphthenate—corrosion mitigation is required to prevent premature fouling due to the buildup of NAC by-products.

4.8 HYDROPROCESSES

Hydroprocesses (hydrogenation processes) are those refining processes in which hydrogen is used in order to control the reaction chemistry and prevent the formation of coke. Hydrogenation processes for the conversion of petroleum fractions and petroleum products may be classified as destructive and nondestructive. Destructive hydrogenation (hydrogenolysis or hydrocracking) is characterized by the conversion of the higher molecular weight constituents in a feedstock to lower boiling products. Such treatment requires severe processing conditions and the use of high hydrogen pressures to minimize polymerization and condensation reactions that lead to coke formation.

Nondestructive or simple hydrogenation is generally used for the purpose of improving product quality without appreciable alteration of the boiling range. Mild processing conditions are employed so that only the more unstable materials are attacked. Nitrogen, sulfur, and oxygen compounds undergo reaction with the hydrogen to remove ammonia, hydrogen sulfide, and water, respectively. Unstable compounds which might lead to the formation of gums, or insoluble materials, are converted to more stable compounds.

Because of the extent of the cracking versus hydrogenation, hydrotreating processes are the least most severe of the hydrogenation processes and can be arbitrarily assigned to be the least corrosive.

4.8.1 Hydrotreating

Catalytic hydrotreating is a hydrogenation process used to remove about 90% w/w of contaminants such as nitrogen, sulfur, oxygen, and metals from liquid petroleum fractions. These contaminants, if not removed from the petroleum fractions as they travel through the refinery processing units, can have detrimental effects on the equipment, the catalysts, and the quality of the finished product. Typically, hydrotreating is done prior to processes such as catalytic reforming so that the catalyst is not contaminated by untreated feedstock. Hydrotreating is also used prior to catalytic cracking to reduce sulfur and improve product yields and to upgrade middle-distillate petroleum fractions into finished kerosene, diesel fuel, and heating fuel oils. In addition, hydrotreating converts olefins and aromatics to saturated compounds.

In the hydrotreating for sulfur removal process (hydrodesulfurization), the feedstock is deaerated and mixed with hydrogen, preheated in a fired heater (315–425°C; 600–800°F), and then charged under pressure (up to 1000 psi) through a fixed-bed catalytic reactor. In the reactor, the sulfur and nitrogen compounds in the feedstock are converted into hydrogen sulfide and ammonia. The reaction products leave the reactor and after cooling to a low temperature enter a liquid/gas separator.

The hydrogen-rich gas from the high-pressure separation is recycled to combine with the feedstock, and the low-pressure gas stream rich in hydrogen sulfide is sent to a gas-treating unit where hydrogen sulfide is removed. The clean gas is then suitable as fuel for the refinery furnaces. The liquid stream is the product from hydrotreating and is normally sent to a stripping column for removal of hydrogen sulfide and other undesirable components. In cases where steam is used for stripping, the product is sent to a vacuum drier for removal of water. Hydrodesulfurized products are blended or used as catalytic reforming feedstock.

4.8.2 Hydrocracking

Hydrocracking is a two-stage process combining catalytic cracking and hydrogenation, wherein heavier feedstocks are cracked in the presence of hydrogen to produce more desirable products. The process employs high pressure, high temperature, a catalyst, and hydrogen. Hydrocracking is used for feedstocks that are difficult to process by either catalytic cracking or reforming, since these feedstocks are characterized usually by a high content of polycyclic aromatic constituents and/or high concentrations of the two principal catalyst poisons, sulfur and nitrogen compounds.

The hydrocracking process largely depends on the nature of the feedstock and the relative rates of the two competing reactions, hydrogenation and cracking. Heavy aromatic feedstock is converted into lighter products under a wide range of very high pressure (1000–2000 psi) and high temperature (400–475°C; 750–1000°F), in the presence of hydrogen and special catalysts. The process is used for feedstocks that are difficult to process by catalytic cracking – these feedstocks are characterized by a high content of polycyclic aromatic constituents and/or high concentrations of sulfur and nitrogen compounds, which poison the catalyst.

In the first stage, preheated feedstock is mixed with recycled hydrogen and sent to the first-stage reactor, where catalysts convert sulfur and nitrogen compounds to hydrogen sulfide and ammonia. Limited hydrocracking also occurs. After the product leaves the first stage, it is cooled and liquefied and run through a hydrocarbon separator. The hydrogen is recycled to the feedstock and the liquid is charged to a fractionator. Depending on the products desired (gasoline components, jet fuel, and gas oil), the fractionator is run to cut out some portion of the first-stage reactor outturn. Kerosene-range material can be taken as a separate side-draw product or included in the fractionator bottoms with the gas oil.

The fractionator bottoms are again mixed with a hydrogen stream and charged to the second stage. Since this material has already been subjected to some hydrogenation, cracking, and reforming in the first stage, the operations of the second stage are more severe (higher temperatures and pressures). Like the outturn of the first stage, the second-stage product is separated from the hydrogen and charged to the fractionator.

Because of the operating temperatures and presence of hydrogen, the hydrogen sulfide content of the feedstock must be strictly controlled to a minimum to reduce the possibility of severe corrosion. Corrosion by wet carbon dioxide in areas of condensation also must be considered. When processing high-nitrogen feedstock, the ammonia and hydrogen sulfide form ammonium hydrosulfide, which causes serious corrosion at temperatures below the water dew point. Ammonium hydrosulfide is also present in sour water stripping.

4.8.3 Effect of Naphthenic Acids
Hydrotreaters are one of the most impacted areas of the refinery as the amount of heavy sour crude increases. Typically, there is significantly more sulfur and aromatics in the products from the atmospheric and vacuum distillation units that need hydroprocessing. Additionally, the coking units will also be producing more distillate and gas oil that also must be hydroprocessed.

With the incremental changes in the feed quality to the unit, such as the blending of high-acid crude oils into the unit feedstock, the demand for hydrogen to achieve the product specifications increases. But, the major concern with the hydroprocessing units and the ability to handle high-acid crudes will be related to the metallurgy. As with the atmospheric and vacuum distillation sections, the NAC can be a major concern while the material is hot. However, many refineries transport the feedstocks between units at moderate temperatures (approximately 205°C, 400°F) that alleviate the need for upgraded interconnecting pipe between units.

However, in the hydroprocessing unit, the feedstock is usually heated as part of the process. Furthermore, once the feedstock is mixed with hydrogen, this prevents 300 series SS alloys (even those without significant molybdenum contents) from experiencing the acidic attack of the naphthenic acid. Therefore, a careful review of the unit configuration is required to ensure that either upgraded metallurgy is provided or that hydrogen is mixed with the oil before heating.

4.9 MITIGATION OF NAC

In order to design and plan a mitigation strategy for any refinery process or unit, it is necessary to know (in this contact, review) the affected units and equipment. Consideration must also be given to piping systems, which are particularly susceptible in areas of high velocity, turbulence, or change of flow direction, such as pump internals, valves, elbows, tees, and reducers as well as areas of flow disturbance such as weld beads and thermowells. The reactor internals may also be corroded in the flash zones, where high-acid streams condense or high-velocity droplets impinge. Furthermore, NAC may also (some might say, is *likely to*) be found in heated feedstock streams downstream of the atmospheric tower and vacuum tower but upstream of any hydroprocessing reactor.

The NAC in any unit or reactor is characterized by localized corrosion, pitting corrosion, or flow-induced grooving in high-velocity areas (Chapters 2 and 3). In low-velocity condensing conditions, many alloys including carbon steel, low-alloy steels, and 400 series stainless steels may show uniform corrosion in terms of loss in thickness and/or pitting (Chapters 2 and 3).

Once the corrosion had been identified and characterized, there are several strategies (related to naphthenic acid constituents — other types of corrosion may require different strategies. The strategies can be implemented after identification of susceptible areas of the unit as well as monitoring the corrosion.

The first part of a strategy to mitigate naphthenic acid corrosion is to carry out a comprehensive analytical assessment of the crude oil which must include a detailed assessment of the naphthenic acid content, the types of naphthenic acids, and the potential for corrosivity (Chapters 1 and 2). This should also include a detailed assessment of the process operating conditions and the character and structure of the processing unit(s). Furthermore, an important part of the strategy is the design and implementation of a comprehensive corrosion-monitoring program. An effective corrosion-monitoring program will help confirm which areas of the unit require a corrosion mitigation strategy, which will provide information about the impact of any necessary corrosion mitigation steps.

The correctly applied strategy should provide (1) a complete understanding of the unit operating conditions, (2) crude oil properties, (3) distillate properties, (4) unit/reactor metallurgy, and (5) equipment performance history thereby allowing a probability of failure analysis to be performed for those areas which would be susceptible to NAC. From the results, each unit/reactor in the processing circuit can be assigned a relative probability failure rating based on the survey data and overall refining industry experience.

Investigations into the appearance of corrosion in the vacuum tower may be due to the occurrence of naphthenic acids or their degradation products in the vacuum tower feedstock. In the vacuum column, preferential vaporization and condensation of naphthenic acids increase the TANs of condensates.

The corrosion is similar to corrosion caused by very high TAN fractions in which it is believed that, in such a case, and velocity may have virtually no effect on the process. The naphthenic acids are most active at their boiling point, but the most severe corrosion generally occurs on condensation. The corrosion mechanism is mainly a condensate corrosion and is directly related to content, molecular weight, and boiling point of the naphthenic acid. Corrosion is typically more severe at the condensing point corresponding to high TAN and temperature. However, in general, vacuum-tower overhead corrosion from naphthenic acid species is normally less severe than atmospheric-tower overhead corrosion because of the large volume of steam that condenses along with the hydrogen in vacuum towers. This steam originates as either bottom-stripping steam or motive steam to the first-stage jet.

In the vacuum tower, preferential vaporization and condensation of naphthenic acids increase the TANs of condensates. There is very little effect of fluid velocity. Corrosion takes place only in the liquid phase. It is mainly a condensate corrosion and is directly related to content, molecular weight, and boiling point of the naphthenic acid.

Corrosion is typically severe at the condensing point corresponding to high TAN and temperature. Simulating vacuum tower corrosion requires the specimens to be exposed in the condensing phase and not in the vapor or liquid phase. This requires apparatus common to corrosion studies in the chemical industry. An autoclave and a flow loop jet impingement are not adequate for testing corrosion in this case.

Also in the vacuum tower, as is the case in the atmospheric tower, preferential vaporization and condensation of naphthenic acids increase the TAN of the condensates. The naphthenic acids are most active at their boiling point, but the most severe corrosion generally occurs where there is condensation of the naphthenic acids (i.e., in the liquid phase). The corrosion mechanism is mainly a condensate corrosion phenomenon and is directly related to naphthenic acid content, molecular weight, and boiling point of the naphthenic acid. Furthermore, corrosion is typically severe at the condensing point corresponding to high TAN and temperature.

For distillation units and/or components of distillation systems (including the fractionators associated with coking, cracking, and hydroprocessing units) that have not been designed for resistance to corrosion by naphthenic acids, the options are to: (1) change or blend crude oils, (2) upgrade the metallurgy of the system, (3) utilize chemical inhibitors, (4) remove the naphthenic acids prior to entry into the refinery proper, that is, after desalting but before distillation or any other form of processing, or (5) some combination thereof (Chapter 5).

Whatever the method chosen, a program of inspection and monitoring must be followed. This should include monitoring the TAN (any other necessary test method) and sulfur content of the crude oil feedstock and side streams to determine the distribution of acids in the distillation fractions from the atmospheric tower and the vacuum tower. This can be (should be) accompanied by ultrasonic testing and radiographic testing for reactor/pipe thickness monitoring. In some cases, localized erosion may be difficult to locate so radiographic testing should be the primary detection method followed by ultrasonic thickness measurement. Electrical resistance corrosion probes and corrosion coupon racks can be used, and streams can be monitored for iron and nickel content to assess corrosion in the system (Speight, 2014b).

Finally, sulfidation is a competing and complimentary mechanism which must be considered in most situations with corrosion due to the presence of naphthenic acids. In cases where thinning is occurring, it is difficult to distinguish between NAC and sulfidation.

The corrosion mechanism at the furnace tubes, transfer lines, areas of high turbulence such as thermowells and pumps is most likely an accelerated corrosion due to the velocity and the two-phase flow. Simulation of these conditions in the laboratory requires conditions of high degree of vaporization and relatively low wall shear stress, especially conditions that bear a strong relationship to the reality of refinery units and refinery operations.

In *side-cut piping*, conditions of low vaporization and medium fluid velocity exist. Some studies showed a possible inhibition of NAC by sulfur compounds. In these conditions, an increase in velocity increases corrosion rates up to the point where impingement starts and corrosion is accelerated dramatically. The corrosion process is dependent on flow, temperature, materials of construction, as well as naphthenic acid content and hydrogen sulfide content.

Process conditions such as load and steam rate and especially turbulence affect corrosivity. The presence of any naphthenic acid most likely increases sulfidic corrosion. The mechanism of corrosion mechanism at the furnace tubes, transfer lines, areas of high turbulence such as thermowells and pumps is most likely due to an accelerated corrosion reaction caused by the high fluid velocity as well as two-phase flow.

In addition to the atmospheric tower and vacuum tower sections, other sections of the refinery susceptible to corrosion include (but may not be limited to) preheat exchanger (caused by hydrogen chloride and hydrogen sulfide), preheat furnace and bottoms exchanger (caused by hydrogen sulfide and sulfur compounds), atmospheric tower and vacuum furnace (caused by hydrogen sulfide, sulfur compounds, and organic acids). Where sour crudes are processed, severe corrosion can occur in furnace tubing and in both atmospheric and vacuum towers where metal temperatures exceed 235°C (450°F). Wet hydrogen sulfide also will cause cracks in steel. When processing high-nitrogen crudes, nitrogen oxides can form in the flue gases of furnaces. Nitrogen oxides (NOx) are corrosive to steel when cooled to low temperatures in the presence of water.

Chemicals are used to control corrosion by hydrochloric acid produced in distillation units. Ammonia may be injected into the overhead stream prior to initial condensation, and/or an alkaline solution may be carefully injected into the hot crude-oil feed. If sufficient wash water is not injected, deposits of ammonium chloride can form and cause serious corrosion. Crude feedstock may contain appreciable amounts of water in suspension which can separate during startup and, along with water remaining in the tower from steam purging, settle in the bottom of the tower. This water can be heated to the boiling point and create an instantaneous vaporization explosion upon contact with the oil in the unit.

CHAPTER 5

Removing Acid Constituents from Crude Oil

5.1 INTRODUCTION

Naphthenic acids having the empirical formula $C_nH_{2n+z}O_2$, occur naturally in crude oil (see Chapter 1) (Clemente and Fedorak, 2005). However, crude oils containing naphthenic acids are only united by the value of the total acid number (TAN) and will vary extensively in most other chemical properties and physical characteristics (see Chapter 1). While most high acid crude oil may be medium to heavy crudes (in terms of the API gravity), such crude oils usually are low in sulfur content (with the notable exception of Venezuelan grades).

Removing naphthenic acid constituents from crude oils and/or preventing acidic corrosion is regarded as one of the most important processes in heavy oil upgrading (Scattergood and Strong, 1987; White and Ehmke, 1991). Current industrial practices either depend on dilution or caustic washing methods to reduce the TAN of heavy crude oils (Table 5.1). However, neither of these approaches is entirely satisfactory.

In addition, process control changes (that are acceptable to the refiner without creating losses in the margin) may provide adequate corrosion control if there is possibility to reduce charge rate and temperature. For long-term reliability, upgrading the construction materials to a higher chrome and/or molybdenum alloy is the best solution. Currently and in the future, modifying or removing naphthenic acids would come from four areas: adsorption, extraction, neutralization (esterification), decarboxylation, and hydrotreating.

However, petroleum research is often frustrated by the wide variability in molecular composition among nominally similar streams. Crude oils vary in composition due to differences in source rock, maturation conditions, and reservoir environments (Speight, 2014a). Process oils are a function of not only the source oil but also process design, catalyst, and operating conditions. Because it is not practical to study every possible compound in a stream, model compounds are used as surrogates for a class of molecules. Typically, tests are run in a benign

Table 5.1 Methods for Mitigating Corrosion due to the Presence of Naphthenic Acids

Blending

- Typically, blend high TAN with low TAN crude
- Blending primarily based on desired product mix
- Metallurgy can become limiting
- Crude compatibility needs evaluation
- Sulfur in blend crude may be critical

Materials Upgrade

- In mild service, 9Cr−1Mo steel is often adequate
- Usually 316L (2% Mo) minimum material
- 317L (3% Mo) often used
- Structured packing requires 317L min.

Use of Inhibitors

- Continuous use of high acid crudes
  - Successful applications exist for wide range of TAN and naphthenic acids
  - Important to maintain monitoring in areas at risk
  - Can be continuous or until metallurgy is upgraded.
- Intermittent use of high acid crudes
  - Used when corrosion rates are excessive based on monitoring
- Cost directly related to amount of equipment protected

matrix oil to facilitate analysis of the compound's behavior. In reality, the research objectives define the type of model compound selected for experiments. Compounds with different functional groups may be used for fundamental physical phenomena or reaction mechanism studies. In process studies, the stream of interest dictates the boiling range and hence a molecular weight range of model compounds to be studied. In sophisticated molecular modeling, molecular structures (isomers) must be considered.

As high acid crude oil and heavy crude oil output rises because of the enhanced exploitation and utilization of oil resources in the world, the acid values of heavy crude oils are also increasing. In recent years, the world's high acid crude oil production has been increasing by 0.3% per year. The acid crude oil can induce serious corrosion of equipment during oil processing, resulting in its oversupply and relatively low price on the international market. The acid value of crude oil is mainly caused by naphthenic acids, which are the major acid components in crude oil, accounting for about 90% of acidic ingredients, which are the main chemical components causing serious corrosion of oil processing equipment (see Chapters 3 and 4). Besides naphthenic acids, there are fatty acids, aromatic acids, inorganic acids, mercaptans, hydrogen sulfide, and phenols existing in the crude oil. Corrosion caused by naphthenic acids is different from that caused by sulfur, because the latter can induce uniform corrosion while the former can cause localized corrosion or pitting, which is influenced by the acid value, temperature, flow rate, kind of medium, changes in physical state, and other factors, it is not always easy to detect the cause of corrosion.

Deacidification of high acid crude oil is considered in two ways: (i) physical methods, such as blending or more obvious physical methods, which are designed to separate the naphthenates from crude oil without changing their chemical state, such as adsorption and extraction and (ii) chemical methods which are associated with taking advantage of chemical properties of the crude oil for naphthenic acid removal such as the use of inhibitors, decarboxylation, and esterification.

Many concepts have been advanced with the goal of removing naphthenic acids (deacidification) from acid crude oils. In developing these processes, in many cases model compounds have been used to probe physical properties and reaction mechanisms. The characteristic of an ideal model compound is defined by research objective, functional group, boiling range, and structure as revealed by advanced characterization of a process stream. However, availability often limits research to less than ideal compounds. This limitation must be recognized in building or evaluating engineering models.

The presence of naphthenic acids in crude oil during refining operations may cause operational issues, such as foaming during crude oil desalting or other operation units as well as carrying cations through the refining process that may cause catalyst deactivation (Shi et al., 2008).

5.2 PHYSICAL METHODS

5.2.1 Blending

Blending (sometimes referred to as *dilution*) may be used to reduce the naphthenic acid content of the feed, thereby reducing corrosion to an acceptable level—the naphthenic acids are not removed but the concentration is reduced through the blending operation. In fact, crude oil blending is the most common solution to high acid crude processing. Blending of higher naphthenic acid content oil with low naphthenic acid content oil can be effective if proper care is taken to control crude oil and distillate acid numbers to proper threshold levels. For example, blending a high acid crude oil with a low acid one may reduce the naphthenic acid content to an acceptable level, but the acidic compounds remain and the value of the low acid oil is diminished. Blending of heavy and light crudes changes shear stress parameters and might also help reduce corrosion. Blending is also used to increase the level of sulfur content in the feed and inhibit to some degree naphthenic acid corrosion.

However, blending two different feedstocks may lead to other issues such as incompatibility of the heavy crude constituents (such as the asphaltene constituents) in the more paraffinic light crude oil (Speight, 2014a). Some oils and refinery streams are inherently incompatible—the main cause being the insolubility of the asphaltene constituents. While there are many models claimed to predict incompatibility, the only true model is based on data from the actual components of the blend under investigation. Rules designed to predict incompatibility are not always adaptable from one system to another and following such rules may not diminish the problems associated with incompatibility.

The common industry practice to overcome the problem of naphthenic acid corrosion consists of blending with sweet crude or washing with caustic solution to lower the acid level, addition of corrosion inhibitor and utilization of expensive corrosion-resistant construction material for the processing unit. Lately, as the production of heavy crude oil continues to increase, these practices have become less satisfactory and other methods of naphthenic acid reduction have been investigated.

Finally, it must be recognized that blending does not deacidify high acid crude oil. The concentration of the naphthenic acids is reduced by blending high acid crude oil to low acid crude oil or no acid crude oil. The naphthenic acids remain in the blended product and those naphthenic acids that have a high propensity to react with a catalyst (resulting in catalyst deactivation or catalyst destruction) will do so.

5.2.2 Adsorption

Clay minerals have and continue to be used in the petroleum industry because they can interact with many organic compounds to form complexes of varying stabilities and properties (Speight and Ozum, 2002; Hsu and Robinson, 2006; Gary et al., 2007; Speight, 2014a). Clay organic interactions are multivariable reactions involving the silicate layers, the inorganic cations, water, and the organic molecules. The chemical affinity between the acid compound and the solid surface depends on structure (molecular weight, chain length, etc.) of the acid molecule, functional groups present in the acid molecule such as hydrophobic groups ($-C-C-C-C-$), electronegative groups ($-C=O$, $-C-O-C-$, $-OH$), π bonds ($-C=C-$, aromatic rings), and configuration of the acid molecule. Adsorption of organic molecules on minerals such as clays has already been considered as a major process in the diagenesis and maturation of organic matter to petroleum. Surface functional groups in clay minerals play a significant role in adsorption processes. Surface functional groups can be organic (e.g., carboxyl, carbonyl, phenolic) or inorganic molecular units. The major inorganic surface functional groups in soils are the siloxane surface associated with the plane of oxygen atoms bound to the silica tetrahedral layer of a phyllosilicate and hydroxyl groups that are associated with the edges of inorganic minerals such as kaolinite, amorphous materials, metal oxides, oxy-hydroxides, and hydroxides.

The presence of the negative surface charge also makes the clay an excellent adsorbent for organic cations. In the present work, it was used as a clay (bentonite) activated with organic acids to increase its surface area. The adsorptive properties of these activated clays depend on the chemical nature of the surface, and the adsorption process is influenced by electrostatic interaction between adsorbate molecules and adsorption sites on clay surface, the nature of the exchangeable ion located in the interlayer, and the range of hydration of the positive ion. Polar molecules or polarizable ones are efficiently adsorbed by these clays. This could favor that acid composites, as naphthenic ones, could be attracted for the negative charge layers of the clay and then removed from oil. These acids can be ionized or bonded in hydrogen bridges inside the oil.

In addition, alumina presents a high surface area for adsorption and its global surface charge is positive, what would favor the adsorption of negatively charged composites, through an electrostatic interaction. This could lead to the assumption that the highest reduction of TANs, obtained with alumina, may result from the adsorption of naphthenic acid totally dissociated. However, as it is a nonaqueous environment, it is not possible to discard the probability that the adsorption process to also occurring through interactions between molecules in the acid form and the weak basic sites present in great amount in alumina. Moreover, it is important to point out that, as this is a real sample, naphthenic acids of differentiated force are possibly present in the oil and can be adsorbed by distinct processes. In addition, other compounds present in the oil can also be adsorbed, contributing to the final TAN.

As these acids are, in general, at low concentrations, an efficient treatment can be the use of adsorption processes. There are few processes reporting the removal of naphthenic acids from petroleum fractions using various adsorbents such as magnesium oxide and aluminum oxide. The adsorption of naphthenic acids has already been reported on zeolites, aluminosilicates from catalyst manufacturing process waste, silica gel, clays, and ion exchange resins. These acids can be recovered using polar solvents (Gaikar and Maiti, 1996; Zou et al., 1997).

Other processes include adsorption using ceramics or clay, or alumina (Pereira Silva et al., 2007; Saad et al., 2014). Adsorption processes using adsorbents such as clay are also in use (Pereira Silva et al., 2007). The adsorption process in clay involves its great surface area and the presence of a negative global charge in its surface, in virtue of the isomorphic replacement of cations in the crystalline net of the mineral. This charge is generally balanced for the adsorption of inorganic cations (e.g., H^+, Na^+, Ca^{2+}) in the internal and external surfaces of material (Gürses et al., 2006).

As another example, the acidity of high acid crude oil (such as Sudanese crude oil) can be reduced by use of clay, which was activated by concentrated sodium hydroxide (Saad et al., 2014). The TAN of Fula high acid crude was 8.51 mg KOH/g, and the TAN of Nile blend was 1.06 mg KOH/g, which can be reduced to 6.27 and 1.05 mg KOH/g, respectively, by activated clay of 3 M NaOH composed mainly of muscovite as explained by XRD analysis. The adsorption naphthenic acids using local activated clay has proved to be an efficient and effective process, but the disposal of spent clay remains an issue.

5.2.3 Extraction

Solvent extraction leads to the generation of excessive quantities of secondary industrial wastewater. Additionally, water/crude emulsions of difficult separation are generated during the process (Ding et al., 2009). *Adsorption* using ionic exchange resins or other adsorbent materials such as clays may only be applied when dealing with light crudes and light distillation fractions.

The most used and effective process to remove naphthenic acids from oils is the liquid–liquid extraction, especially when using ammonia or alkali alcoholic solutions. However, these systems usually form stable emulsions (Gaikar and Maiti, 1996). Therefore, there are several proposals for the liquid–liquid extraction using different solvent systems (Danzik, 1987; Sartori et al., 1997; Varadaraj et al., 1998; Gorbaty et al., 2000; Sartori et al., 2001; Greaney, 2003).

An example of extractive removal of naphthenic acids from crude oil products invokes the use of a solvent system comprising liquid alkanols, water, and ammonia in certain critical ratios to facilitate selective extraction and easy separation. However, when applied to whole crudes, there is the possibility of emulsion formation that will prevent separation of the naphthenic acids (Danzik, 1987).

Naphthenic acids have also been removed from high acid crudes by treating the starting crude oil containing naphthenic acids with an amount of an alkoxylated amine and water under conditions and for a time and at a temperature sufficient to form a water-in-oil emulsion of amine salt to produce a crude oil having decreased amounts of organic acids (Varadaraj et al., 2000).

In another example, a simple solvent system is the sodium hydroxide–ethanol system (Shi et al., 2010). In the proposed process, a sodium hydroxide solution of ethanol was used as the acid removal reagent by mixing with the crude oil and then allowing the two phases to separate, with the naphthenic acids being extracted from the crude oil. Data indicated that the optimal content of sodium hydroxide in crude oil was 3000 μg/g and the optimal extraction time was 5 min with the reagent/oil ratio being 0.4:1 (w/w). The suitable reaction temperature could be room temperature. The TAN of the crude oil was lowered from 3.92 to 0.31 mg KOH/g.

Another potential process (Greaney, 2003) involves contacting a crude oil or a petroleum distillate stream in the presence of an effective amount of water, a base selected from Group IA and Group IIA hydroxides and ammonium hydroxide and a phase transfer agent at an effective temperature (i.e., at which the water is liquid to 180°C, 355°F) for a time sufficient to produce a treated petroleum feed having a decreased naphthenic acid content and an aqueous phase containing naphthenate salts, phase transfer agent, and base. This process facilitates the extraction of higher molecular weight naphthenic acids (in addition to lower molecular weight naphthenic acids), which otherwise would remain in the petroleum stream following extraction with caustic alone. This result is a lower TAN content and, in additional, the presence of the phase transfer agent has been found to reduce the emulsion formation that can occur during caustic treatment leading to enhanced processability.

Examples of suitable phase transfer agents include quaternary onium salts, i.e., basic quaternary onium salts (i.e., hydroxides), nonbasic quaternary onium salts such as quaternary onium halides (e.g., chlorides), hydrogen sulfates, crown ethers, open chain polyethers such as polyethylene glycols. The lengths of the hydrocarbyl chains may be varied within the disclosed ranges and the hydrocarbyl groups may be branched or otherwise substituted with noninterfering groups, provided that the accessibility and suitable organophilic nature are maintained. Also, the structure must allow for close approach and strong electrostatic interaction of the onium cation and the hydroxide anion, OH^-:

$$RCOOH(petroleum) + OH^{-1} \rightarrow RCOO^{-1}(aqueous) + H_2O$$

This resulting anionic species is less soluble in the petroleum stream due to its electrostatic charge and preferentially equilibrates to the aqueous stream.

A new class of solvents, namely the ionic liquids, has recently shown promising application for reducing the acid content in crude oil and these are the ionic liquids, which are reactive solvents (Gordon, 2001). Ionic liquids are a relatively new class of solvents and have shown promising application for reducing the acid content in crude oil. Ionic liquids are comprised of entirely free ions within a liquid state that exist over a wide temperature range.

Being composed entirely of ions, ionic liquids possess negligible vapor pressure, and the wide range of possible cations and anions they contain means that other solvent properties may be easily controlled (Kume et al., 2008). For example, an ionic liquid can be used for deacidification of acidic oil, especially for the removal of the naphthenic acids that can be extracted with the conventional technique. The ionic liquid has demonstrated its good performance for acid removal from the oil sample.

Ionic liquids comprise of entirely free ions within a liquid state that exist over a wide temperature range. Besides reducing the acid content of the crude oil, several research works have also demonstrated the ability of some ionic liquids (imidazolium based) to remove sulfur compounds. In another study, the same ionic liquids with different anions such as thiocyanate, octyl sulfate, and trifluoromethane sulfonate were also shown to be able to extract nitrogen and sulfur compounds (Bosmann et al., 2001; Nie et al., 2006; Hansmeier et al., 2011; Kędra-Krolik et al., 2011).

Thus, the use of ionic liquids could be for a multitude of functions for upgrading the crude oil through removal of the undesirable impurities within a single processing step namely liquid–liquid extraction. In addition, features such as higher thermal stability with extremely low vapor pressure compared to the conventional solvents and coupled with the possibility of regeneration have given significant advantages to ionic liquids for replacing conventional solvents.

In addition to the use of ionic liquids in the petroleum industry, there are already considerable published works discussing the capability of ionic liquids in extracting carboxylic acids investigated the potential of imidazolium ionic liquids as extractants for naphthenic acids (Matsumoto et al., 2004; Marták and Schlosser, 2007).

The ionic liquids 1-n-butyl-imidazolium with three different anions namely thiocyanate (SCN), octyl sulfate (OCS), and trifluoromethane sulfonate have been used to show the potential for this method and extract two types of carboxylic acids namely benzoic acid and n-hexanoic acid, from the hydrocarbon liquid. The results show that the ionic liquids exhibit high extraction efficiency for both carboxylic acids used. Using computational molecular simulation software, the interaction mechanism was investigated based on surface polarization charge densities. From the simulation results, the extraction performance of the ionic liquids can be predicted based on capacity and selectivity parameter (Kamarudin et al., 2012a,b).

However, the application of new technologies such as the use of ionic liquids in the refinery has still to be proven. Many issues have to be solved before, for example, ionic liquids may be successfully applied in the petroleum industry. The economical, technological, and environmental feasibility of large-scale production and utilization of ionic liquids must be asserted before any petroleum company accepts their daily use. The effect of the presence of ionic liquids in crude oil during production, transport, and refining must be asserted in order to identify operational issues. Even if ionic liquids have still to prove their safe utilization in daily and routine petroleum operations, there is a window for the use of these solvents in problems of the petroleum refining industry.

5.3 CHEMICAL METHODS

Naphthenic acids (excluding the mineral acids such as hydrochloric acid) from oil consist primarily of monocarboxylic acids, including aliphatic, naphthenic, and aromatic acids (see Chapter 1). Naphthenic acids are predominantly found in immature heavy crudes due to the fact that they come from the biodegradation in petroleum hydrocarbon reservoirs (Biryukova et al., 2007). The acidity of crude oil is associated with the acid number and it is expressed in milligrams of potassium hydroxide necessary in order to neutralize one gram of crude. Crude oils with acidity levels above 0.5 mg KOH/g are considered potentially corrosive for refinery units (see Chapter 1) (Alvisi and Lins, 2011). Aside from the corrosive effect, naphthenic acids lead to formation of stable emulsions by forming metallic naphthenates that reduce interfacial tension, affecting the processes that involve phase separation stages (Ding et al., 2009). Additionally, calcium naphthenates precipitate along the preheating train and furnaces, promoting coke formation (Simon et al., 2008).

Naphthenic acid removal is one of the most important aspects in safe processing of opportunity crudes. The study and implementation of naphthenic acid removal processes is of vital importance for the adequate exploitation of many heavy types of crude that exhibit high acidity. Investigations regarding acidity reduction in crudes include nondestructive processes such as solvent extraction and adsorption (Gaikar and Maiti, 1996; Wang et al., 2006; Norshahidatul Akmar et al., 2012).

5.3.1 Neutralization
Naphthenic acids have been formerly identified as carboxylic acids in crude oil (see Chapter 1) and it is reasonable to assume that

neutralization (caustic treatment) can be employed to substantially remove naphthenic acid constituents. However, the process generates significant amounts of wastewater and the interfacial properties of the naphthenic acid fraction (see Chapter 1) result in the formation of emulsions that are problematic to treat. In particular, once an emulsion is formed, it is very difficult to remove (Ese et al., 2004).

5.3.2 Use of Inhibitors

Corrosion inhibitors are often the most economical choice for mitigation of naphthenic acid corrosion. Effective inhibition programs can allow refiners to defer or avoid capital intensive alloy upgrades, especially where high acid crudes are not processed on a full-time basis (Duan et al., 2012).

An example of the use of the inhibition of naphthenic acid corrosion includes the use of a polysulfide corrosion inhibitor for inhibiting naphthenic acid corrosion in crude distillation units and furnaces (Petersen et al., 1993). Additionally, a variety of attempts have been made to address the problem of naphthenic acid corrosion by using corrosion inhibitors for the metal surfaces of equipment exposed to the acids, or by neutralizing and removing the acids from the oil. Examples of these technologies include treatment of metal surfaces with corrosion inhibitors such as oil soluble reaction products of an alkyne diol and a polyalkene polyamine (Edmonson, 1987) or by treatment of a liquid hydrocarbon with a dilute aqueous alkaline solution, specifically dilute aqueous NaOH or KOH (Verachtert, 1980). However, issues may arise from the use of aqueous solutions that contain higher concentrations of base because these solutions form emulsions with crude oil, necessitating use of only dilute aqueous base solutions which is the principle behind many alkaline flood recovery operations using alkaline flood (Speight, 2009, 2014a). There is a claim that treatment of heavy oils and petroleum resids having acidic functionalities with a dilute quaternary base such as tetramethylammonium hydroxide in a liquid (alcohol or water) avoids emulsion formation (Liotta, 1981).

Injection of corrosion inhibitors may provide protection for specific fractions that are known to be particularly severe. Monitoring needs to be adequate to assure the effectiveness of the treatment. Process control changes may provide adequate corrosion control if there is the possibility of reducing charge rate and temperature. For long-term reliability, upgrading the construction materials is the best solution. At temperatures above 285°C (550°F), with very low naphthenic acid content, cladding with chromium (Cr) steels (5–12% Cr) is recommended for crudes of >1% sulfur. When hydrogen sulfide is evolved, an alloy containing a minimum of 9% (w/w) chromium is preferred. In contrast to high temperature sulfidic corrosion, low alloy steels containing up to 12% (w/w) Cr do not seem to provide benefits over carbon steel in naphthenic acid service. Type 316 stainless steel (having >2.5% w/w molybdenum, Mo) or Type 317 stainless steel (having >3.5% w/w Mo) is often recommended for cladding of vacuum and atmospheric columns.

While high temperature naphthenic acid corrosion inhibitors have been used with moderate success, potential detrimental effects on downstream catalyst activity must be considered. Inhibitors effectiveness needs to be monitored carefully. For severe conditions, Type 317L stainless steel or other alloys with higher molybdenum content may be required.

5.3.3 Decarboxylation

Thermal decarboxylation and *catalytic decarboxylation* of naphthenic acids are an alternative for the processing of high acidity crudes (Zhang et al., 2004, 2005, 2006; Ding et al., 2009; Yang et al., 2013). Typically, the weak nature of naphthenic acids can be utilized in the neutralization and esterification reactions for converting them into easily removable salts and deacidification of crude oil can also be achieved by destructive hydrogenation for removal of carboxyl radicals directly (Duan et al., 2012).

Thermal decarboxylation can occur during the distillation process (during which the temperature of the crude oil in the distillation column can be as high as 400°C (750°F)). A metal oxide catalyst, magnesium oxide (MgO), has been developed and its effectiveness in catalyzing decarboxylation reactions involving carboxylic acid compounds such as naphthenic acid has been determined based on the formation of carbon dioxide and the conversion of acid (Zhang et al., 2006). The major reaction takes place in the temperature range of 150–300°C (300–570°F):

$$R - CO_2H \rightarrow R - H + CO_2$$

The role of magnesium oxide in the system is considered to be multiple. It has the ability to adsorb acidic compounds via acid—base neutralization and it can also promote reactions such as decarboxylation and hydrocarbon cracking at the increased temperature. Direct application of MgO to crude oil results in significant naphthenic acid removal and lower total acidity of the oil as evidenced by a decrease in RCOOH concentration as determined by Fourier transform infrared spectroscopy (FTIR) and a lower TAN.

Catalytic decarboxylation has also been proven as an effective technique in the removal of naphthenic acids in the crude oil samples using Cu/Mg (10:90) Al_2O_3 and Ni/Mg (10:90) Al_2O_3 catalysts. The calcination temperature for both catalysts was at $1000°C$ ($1830°F$). Both catalysts were highly amorphous (Norshahidatul Akmar et al., 2013).

In another example, Liaohe crude oil with high TAN was subjected to thermal reaction at $300-500°C$ ($570-930°F$). Reaction products were collected and analyzed by negative ion electrospray ionization Fourier transform ion cyclotron resonance mass spectrometry (ESI FT-ICR MS) to determine acid compounds in the crude oil. The double bond equivalence (DBE) versus carbon number was used to characterize the oxygenated components in the feed and reaction products. The O_2 class which mainly corresponds to naphthenic acids decarboxylated at $350-400°C$ ($660-750°F$), resulting in a sharp decrease in the TAN. Phenols (O_1 class) are more thermally stable than carboxylic acids. Carboxylic acids were also thermally cracked into smaller molecular size acids, evidenced by the presence of acetic acid, propanoic acid, and butyric acid in the liquid product, which are also responsible for corrosion problems in refineries (see Chapters 1, 3, 4) (Yang et al., 2013).

The end result of the formation of low molecular weight acidic species is treated in the overheads in refineries. A combined approach to front end treating at crude inlet to heaters and preheat exchangers should be considered. It is commonly assumed that acidity in crude oils is related to carboxylic acid species, i.e., components containing a —COOH functional group. While it is clear that carboxylic acid functionality is an important feature (60% of the ions have two or more oxygen atoms), a major portion (40%) of the acid types are not carboxylic acids. In fact, naphthenic acids are a mixture of different compounds which may be polycyclic and may have unsaturated bonds, aromatic rings, and hydroxyl groups (Rikka, 2007). Even the carboxylic acids are more diverse than expected, with approximately 85% containing more heteroatoms than the two oxygen atoms needed to account for the carboxylic acid groups. Examining the distribution of component types in the acid fraction reveals that there is a broad distribution of species.

Several metal oxide catalysts have been found to be very effective to the catalytic decarboxylation, which were verified by the formation of carbon dioxide. A newly developed catalyst with an additive was able to reduce the TAN of a heavy crude oil from 4.38 to 0.60 at 300°C (570°F) for 4 h. A newly developed catalyst with an additive was able to reduce the TAN of a heavy crude oil from 4.38 to 0.60 at 300°C (570°F) for 4 h. Flow reaction test shows that one of the catalysts we developed can maintain its effectiveness for 12 h at 250°C (480°F). In addition, several natural occurring clays showed promise as adsorption agents to the selective removal of acids. The adsorption capacity of one of the clays was as high as about 70 mg NA/g clay (Zhang et al., 2004).

All of the catalysts (MgO, Ag_2O/Cu_2O, HZSM-5 zeolite, and Pt/Al_2O_3) show excellent catalytic decarboxylation activities at relatively low temperature ca. 200−300°C (390−570°F). The decarboxylation mechanism was investigated through theoretical calculation and product/intermediate analyses. It has been clear that magnesium oxide catalyzes the decarboxylation reaction through a ketone forming mechanism with 2 moles of carboxylic acid, which give rise to the formation of 1 mole of carbon dioxide. Silver oxide (Ag_2O) and cuprous oxide (Cu_2O) are, most possibly, involved in the decarboxylation process via a free radical mechanism. On the other hand, the high activities of zeolite towards decarboxylation would be caused by carbon−carbon bond cracking catalyzed by the strong acidic sites on zeolite. Due to the complexity of the oil composition and poison issues, not all of these catalysts can be directly applied in crude oil. However, their applicability in organic chemistry as well as other functional group modification would be highly predicable.

However, not all acidic species in petroleum are derivatives of carboxylic acids (−COOH) and some of the acidic species are resistant to high temperatures (Speight and Francisco, 1990; Speight, 2014a). For example, acidic species appear in the vacuum residue after having been subjected to the inlet temperatures of an atmospheric distillation tower and a vacuum distillation tower. In addition, for the acid species that are volatile, naphthenic acids are most active at their boiling point and the most severe corrosion generally occurs on condensation from the vapor phase back to the liquid phase.

However, in order to obtain noticeable reduction percentages, it is necessary to operate at temperatures above 250°C (480°F), which could lead to corrosion problems in these units. Esterification is a promising alternative for safe processing of these types of opportunity crudes because it is possible to reach significant acidity reduction percentages at temperatures below 250°C (480°F) even in the absence of a catalyst (Sartori et al., 2001).

The serious problems, such as corrosion and emulsion, might be caused when the crude oil with high naphthenic acid content was processed in refinery. A method was developed by the catalytic esterification process to reduce the acidity of crude oil with high naphthenic acid content. The experimental results demonstrated that ZnMgAl-HTlc was an effective catalyst for the esterification, and structure of ZnMgAl-HTlc had not been changed during the esterification process (Huang et al., 2009). Therefore, this acid removal technique could assist refineries to process the high acid crude oil without upgrading the materials of equipment and pipelines.

Catalytic decarboxylation is a well-established chemical reaction in organic and biochemical processes that has been widely applied in organic synthesis and even applied to the identification of coal structure through oxidative decarboxylation (Ozvatan and Yurum, 2002). Cu-based catalysts, predominately employed homogeneously, are commonly used and, in some cases, the presence of organic nitrogen compounds is also necessary (Darensbourg et al., 1994). Additionally, there are reports that zirconium oxide (ZrO_2) can promote the catalytic decarboxylation of acetic acid in supercritical water. Tungsten complexes facilitate catalytic decarboxylation of cyanoacetic acid through homogeneous catalysis. Zeolite has also been applied in the catalytic decarboxylation of benzoic acid but the reaction occurred at temperatures on the order of 400°C (750°F). Nevertheless, most of these studies are limited to the delicate catalyst system such as transition metal complexes, which have relatively low stabilities at the increased temperatures (Zhang et al., 2004; 2005).

5.3.4 Esterification

Studies concerning acid reduction through esterification have been performed with light distillation currents and obtained satisfactory results using a catalyst such as tin oxide and aluminum oxide ($SnO-Al_2O_3$) (Wang et al., 2007) which can bring about a significant reduction in acidity—1.7 to <0.1 mg KOH/g as measured by the standard test methods for acids in crude oil (see Chapter 1)—using a fixed bed reactor, a methanol/crude ratio of 0.010, and an optimum reaction temperature of 280°C (535°F). Another kinetic study (Wang et al., 2008) of the esterification reaction of naphthenic acids in a diesel fuel used a SnO catalyst upon the reaction's kinetic parameters and showed that the kinetic reactions for esterification must be determined for each type of crude.

5.4 CORROSION MONITORING AND PREVENTION

Combating or preventing corrosion is typically achieved by a complex system of monitoring, preventative repairs, and careful use of materials (Garverick, 1994; Speight, 2014b). In fact, corrosion monitoring is just as important as recognizing the problem and applying controls. Monitoring attempts to assess the useful life of equipment when corrosion conditions change and how effective the controls are. Techniques used for monitoring depend on what the equipment is, what it is used for, and where it is located.

5.4.1 Monitoring and Measurement

Corrosion monitoring techniques can help by: (i) providing an early warning that damaging process conditions exist which may result in a corrosion-induced failure, (ii) studying the correlation of changes in process parameters and their effect on system corrosivity, (iii) diagnosing a particular corrosion problem, identifying its cause and the rate controlling parameters, such as pressure, temperature, pH, and flow rate, (iv) evaluating the effectiveness of a corrosion control/prevention technique such as chemical inhibition and the determination of optimal applications, and (v) providing management information relating to the maintenance requirements and ongoing condition of plant (Speight, 2014b).

Typically, a corrosion measurement, inspection and maintenance program used in any industrial facility will incorporate the measurement elements provided by the four combinations of online/offline, direct/indirect measurements: (i) corrosion monitoring direct, online, (ii) nondestructive testing direct, offline, (iii) analytical chemistry indirect, offline, and (iv) operational data indirect, online. In a well-controlled and coordinated program, data from each source will be used to draw meaningful conclusions about the operational corrosion rates with the process system and how these are most effectively minimized.

Furthermore, analytical testing of process streams is vital to processing high acid crude oils. The monitoring of TANs and other relevant properties is of high importance (Sastri, 1998; Knag, 2005). Tests involving potentiometric titration are normally used for measurement of the TAN. Elements, such as trace metals, should be monitored with inductively coupled plasma (ICP) mass spectrometry or ICP optical emission spectrometry instruments. These machines use ICP for elemental analysis.

Periodic inspections do not, however, deliver continuous pipework condition data that can be correlated with either corrosion drivers or inhibitors to understand the impact of process decisions and the inhibitor usage on plant integrity. Manual acquisition of ultrasonic wall thickness data is also frequently associated with repeatability limitations and data logging errors.

Permanently installed sensor systems, on the other hand, deliver continuous reliable data. The ultrasonic sensors can be installed on pipes and vessels operating at up to 600°C (1110°F)—such sensors have also been certified as safe for use in most hazardous environments. Continuous monitoring through use of appropriate test method data—such as the iron powder test method (Hau et al., 2003) which is used for detecting anomalous cases where oil samples having high acid numbers exhibit less corrosivity than others having much lower acid numbers or where they show completely different corrosivity despite having similar or the same acid number—can validate that when corrosion is occurring it may be an intermittent process rather than a continuous event. In such cases, it is particularly valuable to be able to correlate the data over time with process and/or inhibitor parameters, including fluid dynamics (Cross, 2013).

The field of corrosion measurement, control, and prevention covers a very broad spectrum of technical activities (Speight, 2014b). Corrosion measurement is the quantitative method by which the effectiveness of corrosion control and prevention techniques can be

Table 5.2 Method of Corrosion Measurement	
Nondestructive Testing Analytical Chemistry • Ultrasonic testing • Radiography • Thermography • Eddy current/magnetic flux • Intelligent pigs	Analytical Chemistry • pH measurement • Dissolved gas (O_2, CO_2, H_2S) • Metal ion count (Fe^{2+}, Fe^{3+}) • Microbiological analysis
Operational Data • pH • Flow rate (velocity) • Pressure • Temperature	Fluid Electrochemistry • Potential measurement • Potentiostatic measurements • Potentiodynamic measurements • AC impedance
Corrosion Monitoring • Weight loss coupons • Electrical resistance • Linear polarization • Hydrogen penetration • Galvanic current	

evaluated and provides the feedback to enable corrosion control and prevention methods to be optimized and a wide variety of corrosion measurement techniques exists (Table 5.2).

Some corrosion measurement techniques can be used online, constantly exposed to the process stream, while others provide offline measurement, such as that determined in a laboratory analysis. Some techniques give a direct measure of metal loss or corrosion rate, while others are used to infer that a corrosive environment may exist.

5.4.2 Corrosion Prevention

Corrosion control is an ongoing, dynamic process and in the prevention of metal deterioration by three general ways: (i) change the environment, (ii) change the material, or (iii) place a barrier between the material and its environment. The material does not have to be metal—but *is* in most cases—and the metal does not have to be steel, but, because of the strength and readily availability and cheapness of this material, usually is a metal. Again, the environment is, in most cases, the atmosphere, water, or the earth is an important contributor to corrosion chemistry.

Many of the methods for preventing or reducing corrosion exist, most of them orientated in one way or another toward slowing rates of corrosion and reducing metal deterioration (Bradford, 1993; Jones, 1996). Corrosion control is the prevention of deterioration by mitigating the chemical reactions that cause corrosion in three general ways: (i) change the environment, (ii) change the material, or (iii) place a barrier between the material and its environment. All methods of corrosion control are variations of these general procedures, and many combine more than one of them. The material does not have to be metal but *is* a metal or an alloy of metals in most cases. The metal does not have to be steel, but, because of the strength and cheapness of this material, it usually *is* steel or an alloy of steel. Again, the environment is, in most cases, the atmosphere, water, or the earth, i.e., the constituents of the soil. There are, however, enough exceptions to make corrosion control more complex.

The corrosivity of naphthenic crude oils, its mitigation with chemical inhibitors, and the implementation and continual review of a detailed risk assessment focusing on predicting, monitoring, and mitigating corrosion related effects are essential aspects of processing these crudes (Speight, 2014b).

The critical factors of corrosion by high acid crude oil (or for that matter, any acidic crude oil) must be controlled (Petkova et al., 2009). The corrosion effect of the naphthenic acids can be diminished by blending petroleum of high neutralization number with one of lower neutralization number to obtain raw material with acceptably low acidity. The choice of suitable construction material means the use of corrosion-resistant steel like austenite stainless steel grades— chromium–nickel–molybdenum alloyed ones which have excellent resistance to raw materials containing hydrogen sulfide, chlorides, organic and inorganic acids under high temperature. The introduction of automated corrosion monitoring would also allow the optimization of the amount of inhibitors introduced to change the medium characteristics or technological parameters.

The methods of corrosion control in the presence of naphthenic acids are simply determination of metal loss using corrosion coupons, ultrasonic or hydrogen infiltration measurement of the thicknesses of equipment walls. When supplying petroleum with high content of naphthenic acids, it is necessary to take all precautionary measures before accepting the crude oil for refining. The processing method should be carefully selected to avoid hazardous situations and rapid deterioration of the technological equipment at certain points of the installation for atmospheric distillation of petroleum.

For practical purposes, corrosion in refineries can be classified into low temperature corrosion and high temperature corrosion. Low temperature corrosion is considered to occur below approximately 260°C (500°F) in the presence of water. Carbon steel can be used to handle most hydrocarbon streams in this temperature range, except where aqueous corrosion by inorganic contamination, such as hydrogen chloride or hydrogen sulfide, necessitates selective application of more resistant alloys. High temperature corrosion is considered to take place above approximately 260°C (500°F). The presence of water is not necessary, because corrosion occurs by the direct reaction between metal and environment.

The major cause of low temperature (and, for that matter, high temperature) refinery corrosion is the presence of contaminants in crude oil as it is produced. Although some contaminants are removed during preliminary treating at the wellhead fields as well as during dewatering and desalting it still appears in refinery tankage, along with contaminants picked up in pipelines or marine tankers. However, in most cases, the actual corrosives are formed during initial refinery operations. For example, potentially corrosive hydrogen chloride evolves in crude preheat furnaces from relatively benign calcium chloride ($CaCl_2$) and magnesium chloride ($MgCl_2$) entrained in crude oil (Samuelson, 1954). Mitigation of the problems related to low temperature corrosion is associated with adequate cleaning of the crude oil and removal of corrosive contaminates at the time of, or immediately after, formation.

The pH stabilization technique can be used for corrosion control in wet gas pipelines when no or very little formation water is transported in the pipeline. This technique is based on precipitation of protective corrosion product films on the steel surface by adding pH-stabilizing agents to increase the pH of the water phase in the pipeline. This technique is very well suited for use in pipelines where glycol is used as hydrate preventer, as the pH stabilizer will be regenerated together with the glycol—thus, there is very little need for replenishment of the pH stabilizer.

Some of the ways adapted today to overcome the effects of corrosion are, for example, use of specific types of metal to be longstanding in spite of the effects of corrosion. Carbon steel is used for the majority of refinery equipment requirements as it is cost efficient and withstands most forms of corrosion due to hydrocarbon impurities below a temperature of 205°C (400°F) but as it is not able to resist other chemicals and environments, it is not used universally. Other kinds of metals used are low alloys of steel containing chromium and molybdenum, and stainless steel containing high concentrations of chromium for excessively corrosive environments. More durable metals such as nickel, titanium, and copper alloys are used for the most corrosive areas of the plant which are mostly exposed to the highest of temperatures and the most corrosive of chemicals (Burlov et al., 2013).

Many problems of correct use of corrosion control measures (e.g., injection of chemicals such as inhibitors, neutralizers, biocides, and others) may be solved by means of corrosion monitoring methods (Groysman, 1995, 1996, 1997). For example, hydrocarbons containing water vapors, hydrogen chloride, and hydrogen sulfide leave the atmospheric distillation column at 130°C (265°F). This mixture becomes very corrosive when cooled below the dew point temperature of 100°C (212°F). In order to prevent high acidic corrosion in the air cooler and condensers, neutralizers and corrosion inhibitor are injected in the overhead of the distillation column. In addition, corrosion monitoring equipment should be installed in several places (Speight, 2014b). The more points in the unit used for corrosion monitoring, the better and more efficient is the corrosion coverage.

In summary, control of corrosion requires: (i) evaluation of the potential corrosion risks, (ii) consideration of control options—principally inhibition as well as materials selection, (iii) monitoring whole life cycle suitability, (iv) life cycle costing (LPC) to demonstrate economic choice, and (v) diligent quality assurance (QA) at all stages.

5.4.2.1 Crude Oil Quality

In terms of processing high acid crude oils and mitigating corrosion, crude oil quality is an important aspect of corrosion that is often not recognized as much as other causes (see Chapter 1). Crude oil value—to a refinery—is based on the expected yield and value of the products value, less the operating costs expected to be incurred to achieve the desired yield. Ensuring that the quality of crude oil received is equivalent to the purchased quality (value acquired is equal to value expected) is—with the growing popularity of heavy feedstocks, opportunity crudes, and high acid crudes—one of the greatest challenges facing the refining industry (Cross, 2013; Vetters and Clarida, 2013).

Furthermore, difficulties in minimizing differences between purchased quality and refinery receipt quality are significantly higher when multiple crude oils are processed as a blend, and the complexity of the crude delivery system increases. Shipping crude oil through multiple pipelines and redistribution storage tanks—a reality faced by most inland refiners—results in the delivered crude oil being a composite of the many crude oils. Thus, the resultant composite blend may vary significantly from the expected purchased quality, and the sources of quality problems are much more difficult to estimate.

Issues regarding crude oil properties occur regardless of whether the dominant crude slate is comprised of domestic crude delivered by pipeline or foreign crude delivered via waterborne transportation. In both cases, using simple categories such as gravity or sulfur do not provide an accurate measure of the value of a particular crude oil value, and monitoring only gravity and sulfur does not provide adequate safeguards for the integrity of the crude oil while it is in transit. More sophisticated analyses (with an analysis for constituents likely to cause corrosion and correlated to refinery performance) can provide a comprehensive estimate of quality value to a specific refinery. This analysis needs to give weight to quality consistency, where appropriate, as well as to improved yield, reduced operating expenses, and the compatibility of the crude oil feedstock to the refinery processing hardware.

In addition, a combination of aging plants, greater fluid corrosiveness, and tightening of health, safety, security, and environment requirements has made corrosion management a key consideration for refinery operators. The prevention of corrosion erosion through live monitoring provides a real-time picture of how the refinery is coping with the high demands placed upon it by corrosive fluids. This information can assist in risk management assessments.

5.4.2.2 Acidic Corrosion

Refinery equipment reliability during the processing of high acid crude oils is paramount. Hardware changes—such as upgrading materials construction from carbon steel (CS) and alloy steel to stainless steel (SS) 316/317, which contains molybdenum and is significantly resistant to naphthenic acid corrosion—are complicated tasks and require large capital investment as well as a long turnaround for execution. Alternatives to hardware changes are corrosion mitigation with additives and corrosion monitoring with the application of inspection technologies and analytical tests.

During naphthenic acid crude processing, corrosion at high temperature is mitigated by injecting either phosphate-based ester additives or sulfur-based additives, which provide an adherent layer that does not corrode or erode due to the effect of naphthenic acids. It has been suggested and partially proven that corrosion during processing of high acid crude oils is a lower risk if the sulfur content is high—the relationship between the acid number and amount of sulfur is not fully understood but it does appear that the presence of sulfur-containing constituents has an inhibitive effect (Piehl, 1960; Mottram and Hathaway, 1971; Slavcheva et al., 1998, 1999).

Mitigation of process corrosion includes blending, inhibition, materials upgrading, and process control. Blending may be used to reduce the naphthenic acid content of the feed, thereby reducing corrosion to an acceptable level. Blending of heavy and light crudes can change shear stress parameters and might also help reduce corrosion. Blending is also used to decrease the level of sulfur content in the feed and inhibits, to some degree, naphthenic acid corrosion (see Chapter 1).

5.4.2.3 Sulfidic Corrosion

The presence of sulfur in crude oil can also enhance the corrosive effects of naphthenic acids in the same crude oil (see Chapter 2). Other than carbon and hydrogen, sulfur is the most abundant element in petroleum. It may be present as elemental sulfur, hydrogen sulfide, mercaptans, sulfides, and polysulfides.

However, sulfidic corrosion is differentiated from naphthenic acid corrosion by the corrosion mechanism and the form and structure of the corrosion. While naphthenic acid corrosion is typically characterized as having more localized attack particularly at areas of high velocity and, in some cases, where condensation of concentrated acid vapors can occur in crude distillation units, sulfidic corrosion typically takes the form of a general mass loss or wastage of the exposed surface with the formation of a sulfide corrosion scale.

In addition, the particular forms of sulfur that can participate in this process and the mechanism by which sulfidic corrosion can be understood involves the realization that both sulfur and acid species are present to a varying degree in all crude oils and fractions. In certain limited amounts, sulfur compounds may provide a limited degree of protection from corrosion with the formation of pseudopassivity sulfide films on the metal surfaces. However, increases in either reactive sulfur species or naphthenic acids to levels beyond their threshold limits for various alloys may accelerate corrosion (Kane and Cayard, 2002).

Any of the 18Cr−8Ni, stainless steel grades can be used to control sulfidation. However, it is best to use the stabilized grades mentioned earlier. Some sensitization is unavoidable if exposure in the sensitizing temperature range is continuous or long term. Stainless equipment subjected to such exposure and to sulfidation corrosion should be treated with a 2% (w/w) soda ash solution or an ammonia solution immediately upon shutdown to avoid the formation of polythionic acid which can cause severe intergranular corrosion and stress cracking.

Vessels for high-pressure hydrotreating and other heavy crude fraction upgrading processes (e.g., hydrocracking) are usually constructed of one of the Cr−Mo alloys. To control sulfidation, they are internally clad with one of the 300 series stainless steels by roll or explosion bonding or by weld overlay. In contrast, piping, exchangers, and valves exposed to high temperature hydrogen−hydrogen sulfide environments are usually constructed of solid 300 series stainless alloys. In some designs, Alloy 800H has been used for piping and headers. In others, centrifugally cast HF-modified piping has been used. High nickel alloys are rarely used in refinery or petrochemical plants in hydrogen−hydrogen sulfide environments because of their susceptibility to the formation of deleterious nickel sulfide. They are particularly susceptible to this problem in reducing environments. As a general rule, it is recognized that the higher the nickel in the alloy the more susceptible the material to corrosion.

Vapor diffusion aluminum coatings (*alonizing*) have been used with carbon, Cr−Mo, and stainless steels to help control sulfidation and reduce scaling. For the most part, this has been restricted to smaller components. Aluminum metal spray coatings have also been used but not widely nor very successfully.

5.5 THE FUTURE

Model compounds have been used to screen functional groups, such as naphthenic acids, for effect on corrosion. Given the complexity of the naphthenic acid fraction (see Chapter 1) caution is advised when applying data from model compound studies to the real world of naphthenic acid corrosion. Nevertheless, there are specific conclusions that can be drawn from the work performed that offer some understanding of the chemical and physical effects that play an active role in naphthenic acid corrosion.

In summary, current solutions for mitigating (reducing) naphthenic acid corrosion include (i) blending feedstock to reduce the TAN of the blending material to <1 mg KOH/g, (ii) continuous injection of corrosion inhibitors, and (iii) upgrading material of construction to a higher chrome and/or molybdenum in severely corroded areas of plant. While acid corrosion will still occur, the rate of corrosion will (hopefully) be markedly reduced.

More permanent methods for mitigating naphthenic acid corrosion include destruction of the naphthenic acids by use of: (i) decarboxylation, in which the carboxyl group reacted to produce carbon dioxide—the DOE/California Institute of Technology process employs a catalyst—CaO (2–5 wt% of oil)—at 300°C (570°F) for 4 h which gives a 70% conversion of the naphthenic acids although deactivation of the catalyst due to impurities was a concern, (ii) the Statoil NAR process, which removes naphthenic acid constituents under mild catalytic hydrotreating conditions, (iii) the Exxon process (1998 US Patent) in which 57–88% of the naphthenic acids are destroyed by heating to 400°C (750°F) for 1 h; however, the use of a sweep gas is critical because water in the feed and reaction water needed to be removed, (iv) the Unipure process (1999 US Patent) in which the acid containing feedstock is mixed with lime (CaO), heated to 260°C (500°F), and separated from reacted lime.

Other (nondestructive) methods include (i) the UOP adsorption process UOP (1995 US patent) in which the acids are adsorbed on nickel oxide, (ii) the Exxon process (2001 US patent) in which the naphthenic acids are adsorbed on a strong base ion exchange resin. Patent was only for lube oils and naphthenic acid removal was approximately 50% (w/w), and (iii) the BP extraction process (WO 2000 patent) in which extraction with polar such as methanol (methanol/crude ratio = 1.0) using five extraction stages can lower the TAN from 2.77 to <1.0 mg KOH/g. It is worthy of note at this point that extraction using a simple caustic wash does not suffice as most naphthenic acid constituents (because of their highly hydrocarbon nature—see Chapter 1) have high solubility in the crude oil.

Since high acid crudes are often priced lower than comparable crudes, processing them can have a huge positive impact on the refinery profitability. In our refinery processing, these high acid crudes have resulted in an increased yield of higher value products but some of the realizable value had been offset by the hidden costs and operational difficulty associated with corrosion control programs. These costs negatively impacted the perceived or calculated incremental profit potential, but utilizing new sulfur-based corrosion inhibitor has allowed the refinery to capture the profit gain of higher product yields and controlled corrosion without the operational problems caused by caustic, phosphate ester, and the combined sulfur phosphate ester programs.

Processing high acid crudes which are part of the opportunity crude slate of feedstocks requires higher capital and operating costs than dealing with conventional crude oils. In addition to expanding hydrogen and sulfur plant capacities, processing heavy oil and high acid crudes incurs extra costs because of potential fouling and corrosion problems that lead to poor energy efficiency and increased maintenance requirements. In fact, fouling issues are likely to worsen as refineries process greater volumes of heavier high acid crude. Asphaltene constituents, which make up the highest molecular weight and most polar and aromatic fraction of crude oil, have been blamed for a range of processing problems, including extensive fouling and poor desalter performance (Speight, 2014a,b). Also, high acid crudes are projected to make up approximately 15% of the total crude volume processed worldwide in 2015.

There is also room for the development of alternate methods of naphthenic acid removal from feedstocks. One such method is the use of microwave technology (Huang et al., 2006) and undoubtedly other alternate methods will follow.

The share of high acid crudes in terms of overall crude processed is expected to remain high in the future. These crudes have the potential of offering refiners huge economic benefits due to crude price discounts. However, realizing these benefits requires overcoming the negative impacts of high acidity on product yields and quality, and on the reliability and operations of the refinery.

The rising demand for low sulfur crudes with high gas oil yields will lead to increased imports from acidic crude producers across the globe. The continued need for large volumes of middle distillate will make high acid crudes a common refinery feedstock and since acidic crude worldwide is generally low in sulfur and produced high yields of these products, high acid crudes will be acceptable to many refiners. Unlike other sweet crude oils, high acid crudes will continue to be heavily discounted due to their acidity. Yet acidic crudes will increasingly represent for refiners an ongoing bargain—a group of opportunity crudes, costing less on average, than other available conventional low sulfur crude oils. For example, production of high acid crudes has risen sharply in Asia Pacific, mainly in China, and the parallel growth of the output of high acid crudes in Sudan, West Africa, and Brazil has led many producers to consider accepting such crudes over the long term because of the obvious price differential and the ability to increase product margins.

However, processing high acid crude oil can be risky due to its extreme corrosiveness to processing equipment, which can cause extensive damage, decrease production, and even trigger unexpected refinery outages. In addition, high acid crudes are known to produce diesel with a low cetane number.

In order for refiners to decide whether or not to process high acid crudes, various factors must be considered, with particular consideration given to evolving market conditions and climate change legislation. Since each refinery has its unique set of internal and external challenges and prospects, risk analyses (including strength, weakness, opportunity, and threat) are to be undertaken so that the refiner can maximize profitability (i) by procuring the lower cost crudes, (ii) by making the products in demand both now and in the future, and (iii) by driving down operating costs over the long term.

Naphthenic acid concentration may be a serious source of corrosion for one process and have relatively benign effects for another. However, risk-based assessment of the refinery facilities clearly defines areas for concern if blends of crudes containing naphthenic crudes were to be processed. Analysis will identify all areas exposed to risk of corrosion and requirements to be considered and analyzed prior to processing naphthenic crudes.

In fact, risk-based assessment of the refinery facilities clearly defines areas for concern if blends of crudes containing naphthenic crudes were to be processed (Johnson et al., 2003). Analysis will identify all areas exposed to risk of corrosion and requirements to be considered and analyzed prior to processing naphthenic crudes. High temperature fast flow loop studies can be used more accurately to define the corrosion potential of susceptible areas. Analysis using the fast flow loop provides information on the order and magnitude of corrosive attack by naphthenic species, in the absence of corrosion inhibitors, under conditions anticipated in refinery equipment. It can also be used to conduct metallurgy studies and the effectiveness of corrosion inhibitors applied to the system.

Above all, the uniqueness in process conditions, materials of construction, and blend processed in each refinery and especially the frequent variation in crude or blend processed does still not allow an accurate correlation of plant experience to chemical analysis and laboratory corrosion tests. In addition to corrosion data, there is a need for better monitoring and recording of the process conditions (temperature, velocity, and vaporization) and the analytical data (TAN of cuts, type of acid and sulfur compounds present).

GLOSSARY

Note: This chapter is available on the companion website: http://store. elsevier.com/product.jsp?isbn=9780128006306&_requestid=1050998.

BIBLIOGRAPHY

Note: This chapter is available on the companion website: http://store.
elsevier.com/product.jsp?isbn=9780128006306&_requestid=1050998.

Absolute viscosity a term used interchangeably with viscosity to distinguish it from either kinematic viscosity or commercial viscosity.

Absorbent a material having the power, capacity, or tendency to absorb.

Absorption the assimilation of one material into another; in petroleum refining, the use of an absorptive liquid to selectively remove components from a process stream; a process in which fluid molecules are taken up by a liquid or solid and distributed throughout the body of that liquid or solid.

Accelerated corrosion test method designed to approximate, in a short time, the deteriorating effect under normal long-term service conditions.

Acid any substance containing hydrogen in combination with a non-metal or nonmetallic radical and capable of producing hydrogen ions in solution.

Acid catalyst a catalyst having acidic character; the aluminas are examples of such catalysts.

Acid deposition acid rain; a form of pollution depletion in which pollutants, such as nitrogen oxides and sulfur oxides, are transferred from the atmosphere to soil or water; often referred to as atmospheric self-cleaning. The pollutants usually arise from the use of fossil fuels.

Acid embrittlement a form of hydrogen embrittlement that may be induced in some metals by acid.

Acidity denotes the presence of acid-type constituents whose concentration is usually defined in terms of total acid number. The constituents vary in nature and may or may not markedly influence the behavior of the lubricant.

Acid number the quantity of base (potassium hydroxide) expressed in milligrams of potassium hydroxide that is required to neutralize the acidic constituents in 1 g of sample; a measure of the reactivity of petroleum with a caustic solution and given in terms of milligrams of potassium hydroxide that are neutralized by one gram of petroleum.

Acid rain the precipitation phenomenon that incorporates anthropogenic acids and other acidic chemicals from the atmosphere to the land and water; see Acid deposition.

Acid sludge the residue left after treating petroleum oil with sulfuric acid for the removal of impurities; a black, viscous substance containing the spent acid and impurities.

Acid treating a refining process in which unfinished petroleum products, such as gasoline, kerosene, and lubricating oil stocks, are contacted with sulfuric acid to improve their color, odor, and other properties.

Acid value a measure of acidity. It is normally expressed as milligrams of potassium hydroxide per gram (KOH/g) of sample.

Activated alumina a highly porous material produced from dehydroxylated aluminum hydroxide, is used as a desiccant and as a filtering medium.

Active a state in which a metal tends to corrode (opposite of passive); the negative direction of electrode potential; also used to describe corrosion and is associated with the potential range when an electrode potential is more negative than an adjacent depressed corrosion rate (passive) range; (i) the negative direction of electrode potential; (ii) a state of a metal that is corroding without significant influence of reaction product.

AD atmospheric pressure distillation.

Additive a chemical substance added to a petroleum product to impart or improve certain properties.

Adiabatic a change occurring without loss or gain of heat.

Adsorption adhesion of the molecules of gases, liquids, or dissolved substances to a solid surface, resulting in relatively high concentration of the molecules at the place of contact; e.g., the plating out of an antiwear additive on metal surfaces; transfer of a substance from a solution to the surface of a solid resulting in relatively high concentration of the substance at the place of contact; see also Chromatographic adsorption.

Adsorptive filtration the attraction to, and retention of particles in, a filter medium by electrostatic forces, or by molecular attraction between the particles and the medium.

ADT atmospheric distillation tower; the primary distillation tower of a crude distillation unit which operates at or above atmospheric pressure.

ADU atmospheric distillation unit; generally, a unit for distilling crude at or above atmospheric pressure as opposed to operating under a vacuum.

Aeration the state of air being suspended in a liquid such as a lubricant or hydraulic fluid.

Agglomeration the potential of the system for particle attraction and adhesion.

Alicyclic hydrocarbons compounds containing carbon and hydrogen only which has a cyclic structure (e.g., cyclohexane); also collectively called naphthenes.

Aliphatic hydrocarbons compounds containing carbon and hydrogen only which has an open chain structure (e.g., as ethane, butane, octane, butene) or a cyclic structure (e.g., cyclohexane).

Alkali any substance having basic (as opposed to acidic) properties. In a restricted sense it is applied to the hydroxides of ammonium, lithium, potassium, and sodium. Alkaline materials in lubricating oils neutralize acids to prevent acidic and corrosive wear in internal combustion engines.

Alkaline a high pH usually of an aqueous solution; aqueous solutions of sodium hydroxide, sodium orthosilicate, and sodium carbonate are typical alkaline materials used in enhanced oil recovery.

Alkalinity the capacity of a base to neutralize the hydrogen ion (H^+).

Alkali treatment see Caustic wash.

Alkali wash see Caustic wash.

Alkanes hydrocarbons that contain only single carbon−hydrogen bonds. The chemical name indicates the number of carbon atoms and ends with the suffix "ane."

Alkenes hydrocarbons that contain carbon−carbon double bonds. The chemical name indicates the number of carbon atoms and ends with the suffix "ene."

Alkylation the combination of an unsaturated hydrocarbon (olefin) with a saturated hydrocarbon (paraffin or isoparaffin) to form branched chain saturated hydrocarbons; may also apply to the combination of aromatic hydrocarbons with unsaturated hydrocarbons to form branched chain aromatics.

Alkyl groups a group of carbon and hydrogen atoms that branch from the main carbon chain or ring in a hydrocarbon molecule. The simplest alkyl group, a methyl group, is a carbon atom attached to three hydrogen atoms.

Alumina (Al_2O_3) used in separation methods as an adsorbent and in refining as a catalyst.

Ambient temperature temperature of the area or atmosphere around a process (not the operating temperature of the process itself).

American Petroleum Institute (API) a trade association that promotes US petroleum interests, encourages development of petroleum technology, cooperates with the government in matters of national concern, and provides information on the petroleum industry to the government and the public.

American Society for Testing and Materials (ASTM) a professional society that is responsible for the publication of test methods and the development of test evaluation techniques as well as designing standard test methods for petroleum and other industrial products.

Amine an organic compound containing basic nitrogen; may be toxic and corrosive and the lower molecular weight amines have a smell similar to ammonia.

Amphoteric possession of the quality of reacting either as an acid or as a base.

Anaerobic free of air or uncombined oxygen.

Anhydrous the absence of water.

Aniline point the minimum temperature for complete miscibility of equal volumes of aniline and the sample under test (ASTM D611). A product of high aniline point will be low in aromatics and naphthenes and, therefore, high in paraffins. Aniline point is often specified for spray oils, cleaning solvents, and thinners, where effectiveness depends upon aromatic content. In conjunction with API gravity, the aniline point may be used to calculate the net heat of combustion for aviation fuels.

Antifouling preventing fouling; see Fouling.

Atmospheric corrosion the gradual degradation or alterations of a material by contact with substances present in the atmosphere, such as oxygen, carbon dioxide, water vapor, and sulfur and chlorine compounds.

API American Petroleum Institute—a trade association of petroleum producers, refiners, marketers, and transporters, organized for the advancement of the petroleum industry by conducting research, gathering and disseminating information, and maintaining cooperation between government and the industry on all matters of mutual interest.

API gravity a gravity scale established by the American Petroleum Institute and in general use in the petroleum industry; a measure of the *lightness* or *heaviness* of petroleum which is related to density and specific gravity.

$$API = (141.5/sp\ gr@60°F) - 131.5$$

Apparent viscosity the ratio of shear stress to rate of shear of a non-Newtonian fluid such as lubricating grease, or a multi-grade oil, calculated from Poiseuille's equation and measured in poises; the apparent viscosity changes with changing rates of shear and temperature and must, therefore, be reported as the value at a given shear rate and temperature (ASTM D1092).

Aromatic derived from, or characterized by, the presence of the benzene ring.

Aromatic hydrocarbon a hydrocarbon characterized by the presence of an aromatic ring or condensed aromatic rings; benzene and substituted benzene, naphthalene and substituted naphthalene, phenanthrene and substituted phenanthrene, as well as the higher condensed ring systems; compounds that are distinct from those of aliphatic compounds (*q.v.*) or alicyclic compounds (*q.v.*).

Aromatization the conversion of nonaromatic hydrocarbons to aromatic hydrocarbons by: (i) rearrangement of aliphatic (noncyclic) hydrocarbons (*q.v.*) into aromatic ring structures and (ii) dehydrogenation of alicyclic hydrocarbons (naphthenes).

Ash A measure of the amount of inorganic material in petroleum or a petroleum fraction or petroleum product; determined by burning the sample in air weighing the residue; the results expressed as percent by weight.

Ash content erroneous name for the ash produced from the mineral matter content of petroleum or a petroleum derived product—petroleum and petroleum products do not contain ash; more correctly, *ash yield* which is the percent by weight of residue remaining after combustion of a sample of petroleum.

Asphalt the nonvolatile product obtained by distillation and treatment of an asphaltic crude oil; a manufactured product.

Asphalt cement asphalt especially prepared as to quality and consistency for direct use in the manufacture of bituminous pavements.

Asphalt emulsion an emulsion of asphalt cement in water containing a small amount of emulsifying agent.

Asphalt flux an oil used to reduce the consistency or viscosity of hard asphalt to the point required for use.

Asphalt primer a liquid asphaltic material of low viscosity which upon application to a nonbituminous surface to waterproof the surface and prepare it for further construction.

Asphaltene (asphaltene constituents) the brown-to-black powdery material produced by treatment of petroleum, heavy oil, petroleum residua, or bituminous materials with a low boiling liquid hydrocarbon, e.g., *n*-pentane or *n*-heptane; soluble in benzene (and other aromatic solvents), carbon disulfide, and chloroform (or other chlorinated hydrocarbon solvents).

ASTM American Society for Testing Materials; a society for developing standards for testing petroleum and petroleum products; now known as ASTM International.

ASTM International see ASTM.

Atm atmosphere.

Atmospheric pressure pressure exerted by the atmosphere at any specific location. (Sea level pressure is approximately 14.7 pounds per square inch absolute.)

Atmospheric residuum a residuum (*q.v.*) obtained by distillation of a crude oil under atmospheric pressure and which boils above 350°C (660°F).

Atmospheric equivalent boiling point (AEBP) a mathematical method of estimating the boiling point at atmospheric pressure of nonvolatile fractions of petroleum.

Attapulgus clay see Fuller's earth.

Austenite a solid solution of one or more elements in face-centered cubic iron; unless otherwise designated (such as nickel austenite), the solute is generally assumed to be carbon.

Austenitic steel steel with a microstructure at room temperature that consists predominantly of austenite.

Barrel a unit of liquid volume of petroleum oils equal to 42 US gallons or approximately 35 imperial (UK) gallons.

Base a material which neutralizes acids. An oil additive containing colloidally dispersed metal carbonate, used to reduce corrosive wear.

Base number the amount of acid, expressed in terms of the equivalent number of milligrams of potassium hydroxide, required to neutralize all basic constituents present in 1 g of sample.

Basic nitrogen nitrogen (in petroleum) which occurs in pyridine form.

Basic sediment and water (bs&w, bsw) the material which collects in the bottom of storage tanks, usually composed of oil, water, and foreign matter; also called bottoms, bottom settlings.

Bbl. barrel.

Bell cap a hemispherical or triangular cover placed over the riser in a (distillation) tower to direct the vapors through the liquid layer on the tray; see Bubble cap.

Bentonite montmorillonite (a magnesium—aluminum silicate); used as a treating agent.

Benzene a colorless aromatic liquid hydrocarbon (C_6H_6).

BFOE barrels fuel oil equivalent based on net heating value of 6,050,000 BTU per BFOE.

Billion 1×10^9.

Biocide any chemical capable of killing bacteria and bioorganisms.

Biocorrosion biocorrosion processes at metal surfaces are associated with microorganisms, or the products of their metabolic activities including enzymes, exopolymers, organic and inorganic acids, as well as volatile compounds such as ammonia or hydrogen sulfide. These can affect cathodic and/or anodic reactions, thus altering electrochemistry at the biofilm/metal interface; see Microbial corrosion, Microbiologically influenced corrosion.

Biogenic sulfide corrosion the process of forming hydrogen sulfide gas and the subsequent conversion to sulfuric acid (in the presence of moisture) which then causes corrosion of metal pipelines; see Microbial corrosion.

Biological corrosion deterioration of metals as a result of the metabolic activity of microorganisms.

Biomass biological organic matter.

Bitumen also (incorrectly) called asphalt, pitch, or tar; occurs in nature as asphalt lakes (such as the Trinidad Asphalt Lake) and tar sands (such as the Athabasca Tar Sands or Athabasca Oil Sands in Alberta, Canada); consists of high molecular weight hydrocarbonaceous compounds which contain sulfur and nitrogen compounds.

Bituminous containing bitumen or constituting the source of bitumen.

Bituminous rock see Bituminous sand.

Bituminous sand a formation in which the bituminous material (see Bitumen) is found as a filling in veins and fissures in fractured rock or impregnating relatively shallow sand, sandstone, and limestone strata; a sandstone reservoir that is impregnated with a heavy, viscous black petroleum-like material that cannot be retrieved through a well by conventional production techniques.

Black oil a name applied to any oil that is dark colored; generally an indistinct and meaningless term that cannot be applied to the nomenclature and/or classification of petroleum.

Blister a raised area, often dome shaped, resulting from (i) loss of adhesion between a coating or deposit and the base metal or (ii) delamination under the pressure of expanding gas trapped in a metal in a near-subsurface zone; very small blisters are known as *pinhead blisters* or *pepper blisters*.

Blushing whitening and loss of gloss of a coating, usually organic, caused by moisture (also known as *blooming*).

Boiling point the temperature at which a substance boils or is converted into vapor by bubbles forming within the liquid; it varies with pressure.

Boiling range for a mixture of substances, such as a petroleum fraction, the temperature interval between the initial and final boiling points.

Bottoms the nonvolatile portion of crude oil, usually referred to as *residuum*; the liquid which collects in the bottom of a vessel (tower bottoms, tank bottoms) during distillation; also the deposit or sediment formed during storage of petroleum or a petroleum product; see also Residuum and Basic sediment and water.

Brine seawater containing a higher concentration of dissolved salt than that of the ordinary ocean.

British thermal unit see BTU.

Brittle fracture fracture with little or no plastic deformation.

Bromine number a test which indicates the degree of unsaturation in the test sample.

BS&W bottom sediment and water; the heavy material which collects in the bottom of storage tanks; composed of oil, water, and foreign matter.

BSW see BS&W.

British thermal unit (BTU) the amount of heat required to raise the temperature of 1 pound of water 1 degree Fahrenheit.

BTU see British thermal unit.

Bubble cap an inverted cup with a notched or slotted periphery to disperse the vapor in small bubbles beneath the surface of the liquid on the bubble plate in a distillation tower.

Bubble plate a tray in a distillation tower.

Bubble point the temperature at which incipient vaporization of a liquid in a liquid mixture occurs, corresponding with the equilibrium point of 0% vaporization or 100% condensation.

Bubble tower a fractionating tower so constructed that the vapors rising pass up through layers of condensate on a series of plates or trays (see Bubble plate); the vapor passes from one plate to the next above by bubbling under one or more caps (see Bubble cap) and out through the liquid on the plate where the less volatile portions of vapor condense in bubbling through the liquid on the plate, overflow to the next lower plate, and ultimately back into the reboiler thereby effecting fractionation.

Bubble tray a circular perforated plate having the internal diameter of a bubble tower (*q.v.*), set at specified distances in a tower to collect the various fractions produced during distillation.

Bunker C oil see No. 6 fuel oil.

Burner fuel oil any petroleum liquid suitable for combustion.

C_1, C_2, C_3, C_4, C_5 fractions a common way of representing fractions containing a preponderance of hydrocarbons having 1, 2, 3, 4, or 5 carbon atoms, respectively, and without reference to hydrocarbon type.

C or cent. centigrade or celsius.

Carbon residue the amount of carbonaceous (coke-like) residue remaining after thermal decomposition of petroleum, a petroleum fraction, or a petroleum product in a limited amount of air; also called the *coke-* or *carbon-forming propensity*; often prefixed by the terms Conradson or Ramsbottom in reference to the inventor of the respective tests; often called *thermal coke.*

Carbon type the distinction between paraffinic, naphthenic, and aromatic molecules. In relation to lubricant base stocks, the predominant type present.

Cascade tray a fractionating device consisting of a series of parallel troughs arranged on stair-step fashion in which liquid from the tray above enters the uppermost trough and liquid thrown from this trough by vapor rising from the tray below impinges against a plate and a perforated baffle and liquid passing through the baffle enters the next longer of the troughs.

Catalyst a substance that initiates or increases the rate of a chemical reaction, without itself being used up in the process; a chemical substance, usually present in small amounts relative to the reactants, that increases the rate at which a chemical reaction (e.g., corrosion) would otherwise occur, but is not consumed in the reaction.

Catalyst selectivity the relative activity of a catalyst with respect to a particular compound in a mixture, or the relative rate in competing reactions of a single reactant.

Catalytic cracking the conversion of high boiling feedstocks into lower boiling products by means of a catalyst which may be used in a fixed bed (*q.v.*) or fluid bed (*q.v.*).

Cat cracking see Catalytic cracking.

Catalytic dewaxing a catalytic hydrocracking process which uses catalysts such as molecular sieves to selectively hydrocrack the waxes present in hydrocarbon fractions.

Caustic (i) burning or corrosive, (ii) a hydroxide of a light metal, such as sodium hydroxide (NaOH) or potassium hydroxide (KOH).

Caustic wash the process of treating a product with a solution of caustic soda to remove minor impurities; often used in reference to the solution itself.

Cetane index an approximation of the cetane number (*q.v.*) calculated from the density (*q.v.*) and mid-boiling point temperature (*q.v.*); see also Diesel index.

Cetane number a number indicating the ignition quality of diesel fuel; a high cetane number represents a short ignition delay time; the ignition quality of diesel.

Centipoise (cP) a unit of absolute viscosity. 1 centipoise = 0.01 poise.

Centistoke (cSt) a unit of kinematic viscosity. 1 centistoke = 0.01 stoke.

Characterization factor an index of feedstock quality used for correlating data based on physical properties; the Watson (UOP) characterization factor is defined as the cube root of the mean average boiling point in degrees Rankine divided by the specific gravity.

Chevron pattern a V-shaped pattern on a fatigue or brittle fracture surface—the pattern can also be one of straight radial lines on cylindrical specimens.

Chloride stress corrosion cracking cracking of a metal under the combined action of tensile stress and corrosion in the presence of chlorides and an electrolyte (usually water).

Chromatographic adsorption selective adsorption on materials such as activated carbon, alumina, or silica gel; liquid or gaseous mixtures of hydrocarbons are passed through the adsorbent in a stream of diluent, and certain components are preferentially adsorbed.

Chromatographic separation the separation of different species of compounds according to their size and interaction with the rock as they flow through a porous medium.

Chromatography a method of separation based on selective adsorption; see also Chromatographic adsorption.

Clad metal a composite metal containing two or more layers that have been bonded together; the bonding may have been accomplished by co-rolling, co-extrusion, welding, diffusion bonding, casting, heavy chemical deposition, or heavy electroplating.

Clay silicate minerals that also usually contain aluminum and have particle sizes <0.002 micron; used in separation methods as an adsorbent and in refining as a catalyst.

Clay filtration a refining process using fuller's earth (activated clay), bauxite, or other mineral to absorb minute solids from lubricating oil, as well as remove traces of water, acids, and polar compounds.

Clay refining a treating process in which vaporized gasoline or other light petroleum product is passed through a bed of granular clay such as fuller's earth (*q.v.*).

Clay regeneration a process in which spent coarse-grained adsorbent clays from percolation processes are cleaned for reuse by deoiling them with naphtha, steaming out the excess naphtha, and then roasting in a stream of air to remove carbonaceous matter.

Clay treating a clay adsorption process operated at elevated temperature and pressure used to neutralize or improve the color and stability of a lube base oil.

Cleveland open cup a flash point test in which the surface of the sample is completely open to the atmosphere, and which is therefore relatively insensitive to small traces of volatile contaminants.

Cloud point the temperature at which waxy crystals in an oil or fuel form a cloudy appearance.

Coke a gray to black solid carbonaceous material produced from petroleum during thermal processing; characterized by having a high carbon content (95%+ by weight) and a honeycomb type of appearance and is insoluble in organic solvents.

Coke drum a vessel in which coke is formed and which can be cut from the drum prior to drum cleaning.

Coke number used, particularly in Great Britain, to report the results of the Ramsbottom carbon residue test (*q.v.*), which is also referred to as a coke test.

Coker the processing unit in which coking takes place.

Coking a process for the thermal conversion of petroleum in which gaseous, liquid, and solid (coke) products are formed; the undesirable accumulation of carbon (coke) deposits in a refinery; the process of distilling a petroleum product to dryness.

Condensate a mixture of light hydrocarbon liquids obtained by condensation of hydrocarbon vapors predominately butane, propane, and pentane with some heavier hydrocarbons and relatively little methane or ethane; see also Natural gas liquids.

Conradson carbon residue the residue remaining as the result of a test method used to determine the amount of carbon residue left after the evaporation and pyrolysis of the test sample at specified conditions.

Contaminant any foreign or unwanted substance that can have a negative effect on system operation, life, or reliability.

Conversion the thermal treatment of petroleum which results in the formation of new products by the alteration of the original constituents.

Conversion factor the percentage of feedstock converted to light ends, gasoline, other liquid fuels, and coke.

Copper strip corrosion the gradual eating away of copper surfaces as the result of oxidation or other chemical action. It is caused by acids or other corrosive agents.

Corrosion the irreversible chemical deterioration of a material, usually a metal or a metal alloy, because of a reaction with its environment; the name for chemical erosion; the disintegration of an engineered material into its constituent atoms due to chemical reactions with its surroundings; commonly this is the electrochemical oxidation of metals in reaction with an oxidant such as oxygen—the formation of an oxide of iron due to oxidation of the iron atoms in solid solution is a well-known example of electrochemical corrosion, commonly known as *rusting*.

Corrosion control the measures used to prevent or considerably reduce the effects of corrosion. Some practices for corrosion control involve cathodic protection, chemical inhibition, chemical control (removal of dissolved gases such as hydrogen sulfide, carbon dioxide, and oxygen), oxygen scavenging, pH adjustment, deposition control (e.g., scales), and coatings. The corrosion rate will vary with time depending on the particular conditions to which the pipeline is subjected, such as the amount of water present and pressure variations; corrosion control is a continuous process in pipeline operations.

Corrosion fatigue fatigue-type cracking of metal caused by repeated or fluctuating stresses in a corrosive environment characterized by shorter life than would be encountered as a result of either the repeated or fluctuating stress alone or the corrosive environment alone.

Corrosion inhibitor additive for protecting lubricated metal surfaces against chemical attack by water or other contaminants; there are several types of corrosion inhibitors. Polar compounds wet the metal surface preferentially, protecting it with a film of oil.

Corrosion protection Every measure which serves to reduce or prevent corrosion damages is called corrosion protection. In surface technology, protective coatings, e.g., paint/lacquer or metallic top coats, are applied frequently.

Corrosion rate the amount of corrosion occurring in unit time—for example, mass change per unit area per unit time; penetration per unit time; the weight loss of a corrosion coupon after exposure to a corrosive environment, expressed as mils (thousandths of an inch) per year penetration. Corrosion rate is calculated assuming uniform corrosion over the entire surface of the coupon.

mpy = (weight loss in grams) $\times$ (22,300)/(ADT)

mpy is corrosion rate (mils per year penetration)—a mil is one thousandth of an inch.

A is area of coupon (sq. in.)

D is metal density of coupon (g/cm^3)

T is time of exposure in corrosive environment (days).

It is important to note that the calculated values using this formula are not representative in cases of severe pitting. Therefore, a complete report, including a visual inspection, is required to determine either the type of attack or the appropriate corrosion control program. Corrosion rate is also known as corrosion ratio.

Corrosion resistance ability of a material, usually a metal, to withstand corrosion in a given system.

Corrosion resistant alloy a specially formulated material used for components in pipelines likely to present corrosion problems.

Corrosiveness (corrosivity) the tendency of an environment to cause corrosion.

Counterpoise a conductor or system of conductors arranged beneath a power line, located on, above, or most frequently, below the surface of the earth and connected to the footings of the towers or poles supporting the power line.

Coupon a specimen of material exposed to tests or real environments to assess the effect of degradation on the material.

cP centipoise, unit of dynamic viscosity.

Cracked residua residua that have been subjected to temperatures above 350°C (660°F) during the distillation process.

Cracking the process whereby large molecules are broken down by the application of heat and pressure to form smaller molecules.

Cracking activity see Catalytic activity.

Cracking pressure the pressure at which a pressure operated valve begins to pass fluid.

Cracking temperature the temperature (350°C; 660°F) at which the rate of thermal decomposition of petroleum constituents becomes significant.

Crevice corrosion caused by a difference of oxygen availability between two sites on a passive metal that leads to the formation of an electrochemical cell—a selective attack within cracks and at other sites of poor oxygen access is frequently observed; localized corrosion of a metal surface at, or immediately adjacent to, an area that is shielded from full exposure to the environment because of close proximity between the metal and the surface of another material. Chemical deterioration of a material, usually a metal, because of a reaction with its environment.

Crude oil see Petroleum.

Crude still distillation (*q.v.*) equipment in which crude oil is separated into various products.

cSt centistokes, unit of kinematic viscosity.

Cut the portion or fraction of a crude oil boiling within certain temperature limits.

Cut point the temperature limit of a cut or fraction, usually but not limited to a true boiling point basis; the boiling temperature division between distillation fractions of petroleum.

Deactivation reduction in catalyst activity by the deposition of contaminants (e.g., coke, metals) during a process.

Deaerator a separator that removes air from the system fluid through the application of bubble dynamics.

Dealloying the selective corrosion of one or more components of a solid solution alloy (also known as *parting* or *selective dissolution*).

Deasphalted oil the extract or residual oil from which asphaltene and resin constituents have been removed by an extractive precipitation process called deasphalting; typically the soluble material after the insoluble asphaltic constituents have been removed; commonly, but often incorrectly, used in place of *deasphaltened oil*; see Deasphalting.

Deasphaltened oil the fraction of petroleum after only the asphaltene constituents have been removed.

Deasphaltening removal of a solid powdery asphaltene fraction from petroleum by the addition of the low boiling liquid hydrocarbons such as *n*-pentane or *n*-heptane under ambient conditions.

Deasphalting the removal of the asphaltene fraction from petroleum by the addition of a low boiling hydrocarbon liquid such as *n*-pentane or *n*-heptane; more correctly the removal asphalt (tacky, semisolid) from petroleum (as occurs in a refinery asphalt plant) by the addition of liquid propane or liquid butane under pressure; also, a commercial process for removing asphalt from reduced crude or vacuum residua (residual oil) which utilizes the different solubility of asphaltic and nonasphaltic constituents in low boiling hydrocarbon liquids, e.g., liquid propane.

Dehydrator a separator that removes water from the petroleum system.

Delayed coking a coking process in which the thermal reaction is allowed to proceed to completion to produce gaseous, liquid, and solid (coke) products.

Density the mass of a unit volume of a substance; the numerical value varies with the units used.

Desalting removal of mineral salts (mostly chlorides) from crude oils.

Desorption the converse of absorption or adsorption; in filtration, it relates to the downstream release of particles previously retained by the filter.

Desulfurization the removal of sulfur or sulfur compounds from a feedstock.

Dewaxing removal of wax from a petroleum product in order to reduce the pour point; solvent dewaxing is the process in which a number of different solvents can be used has the following steps: feedstock is mixed with solvent and chilled; wax precipitated from solution is separated; solvent is recovered from wax and dewaxed oil; wax separation is accomplished by filtration, centrifuging, or settling; see Solvent dewaxing.

Dew point pressure the pressure at which the first drop of liquid is formed, when it goes from the vapor phase to the two-phase region.

Dew point temperature the temperature at which the first drop of liquid is formed, when it goes from the vapor phase to the two-phase region.

Dissolved gases gases that enter into solution with a fluid and are neither free nor entrained gases.

Dissolved water water which is dispersed in the fluid to form a mixture.

Distillation a process for separating liquids with different boiling points.

Distillation curve see Distillation profile.

Distillation loss the difference, in a laboratory distillation, between the volume of liquid originally introduced into the distilling flask and the sum of the residue and the condensate recovered.

Distillation Method (ASTM D95) A method involving distilling the fluid sample in the presence of a solvent that is miscible in the sample but immiscible in water. The water distilled from the fluid is condensed and segregated in a specially designed receiving tube or tray graduated to directly indicate the volume of water distilled.

Distillation range the difference between the temperature at the initial boiling point and at the end point, as obtained by the distillation test.

Distillation profile the distillation characteristics of petroleum or petroleum products showing the temperature and the percent distilled.

Domestic heating oil see No. 2 fuel oil.

Downcomer a means of conveying liquid from one tray to the next below in a bubble tray column (*q.v.*).

Electric desalting a continuous process to remove inorganic salts and other impurities from crude oil by settling out in an electrostatic field.

Electrical precipitation a process using an electrical field to improve the separation of hydrocarbon reagent dispersions. May be used in chemical treating processes on a wide variety of refinery stocks.

Electrochemical reaction any chemical transformation that implies the transfer of charge across the interface between an electronic conductor (the electrode) and an ionic conductor (the electrolyte).

Electrostatic precipitators devices used to trap fine dust particles (usually in the size range 30−60 microns) that operate on the principle of imparting an electric charge to particles in an incoming air stream and which are then collected on an oppositely charged plate across a high-voltage field.

Electrostatic separator a separator that removes contaminant from dielectric fluids by applying an electrical charge to the contaminant that is then attracted to a collection device of different electrical charge.

Embrittlement steel and other metals and alloys can be embrittled by environmental conditions (environmentally assisted embrittlement); the forms of environmental embrittlement include acid embrittlement, caustic embrittlement, corrosion embrittlement, creep-rupture embrittlement, hydrogen embrittlement, liquid embrittlement, neutron embrittlement, solder embrittlement, solid metal embrittlement, and stress corrosion cracking.

Emulsion intimate mixture of oil and water, generally of a milky or cloudy appearance.

Emulsions may be of two types oil-in-water (where water is the continuous phase) and water-in-oil (where water is the discontinuous phase).

Emulsion breaking the settling or aggregation of colloidal-sized emulsions from suspension in a liquid medium.

Environment the surroundings or conditions (physical, chemical, mechanical) in which a material exists.

EP End point, usually end point of a distillation process.

EPA United States Environmental Protection Agency.

Erosion the progressive removal of a machine surface by cavitation or by particle impingement at high velocities; abrasive metal loss caused by high surface velocity of the transported media, particularly when entrained solids or particulates are present.

Erosion corrosion the result of an electrochemical reaction combined with a material loss by mechanical wear due to impingement of solids or a fluid; corrosion which is increased because of the abrasive action of a moving stream; the presence of suspended particles greatly accelerates abrasive action.

Exfoliation a thick, layer-like growth of loose corrosion products (observed in some cases on steel and aluminum alloys).

Exfoliation corrosion localized subsurface corrosion in zones parallel to the surface that result in thin layers of uncorroded metal resembling the pages of a book.

Expanding clays clays that expand or swell on contact with water, e.g., montmorillonite.

Extract the portion of a sample preferentially dissolved by the solvent and recovered by physically separating the solvent.

Extractive distillation the separation of different components of mixtures which have similar vapor pressures by flowing a relatively high boiling solvent, which is selective for one of the components in the feed, down a distillation column as the distillation proceeds; the selective solvent scrubs the soluble component from the vapor.

Extra heavy oil crude oil with relatively high fractions of heavy components, high specific gravity (low API density), and high viscosity but mobile at reservoir conditions; thermal recovery methods are the most common form of commercially exploiting this kind of oil.

FCC fluid catalytic cracking.

FCCU fluid catalytic cracking unit.

Feedstock petroleum as it is fed to the refinery; a refinery product that is used as the raw material for another process; the term is also generally applied to raw materials used in other industrial processes.

Filiform corrosion corrosion that occurs under a coating in the form of randomly distributed thread-like filaments.

Filter any device or porous substance used as a strainer for cleaning fluids by removing suspended matter.

Filter efficiency method of expressing a filter's ability to trap and retain contaminants of a given size.

Filtration the physical or mechanical process of separating insoluble particulate matter from a fluid, such as air or liquid, by passing the fluid through a filter medium that will not allow the particulates to pass through it.

Fire point (Cleveland open cup) the temperature to which a combustible liquid must be heated so that the released vapor will burn continuously when ignited under specified conditions.

Fixed bed a stationary bed (of catalyst) to accomplish a process (see Fluid bed).

Flash point (Cleveland open cup) the temperature to which a combustible liquid must be heated to give off sufficient vapor to form momentarily a flammable mixture with air when a small flame is applied under specified conditions (ASTM D92).

Flexicoking a modification of the fluid coking process insofar as the process also includes a gasifier adjoining the burner/regenerator to convert excess coke to a clean fuel gas.

Flocculation threshold the point at which constituents of a solution (e.g., asphaltene constituents or coke precursors) will separate from the solution as a separate (solid) phase.

Floc point (flocculation point) the temperature at which wax or solids separate in oil.

Fluid bed a bed (of catalyst) that is agitated by an upward passing gas in such a manner that the particles of the bed simulate the movement of a fluid and has the characteristics associated with a true liquid; c.f. Fixed bed.

Fluid catalytic cracking cracking in the presence of a fluidized bed of catalyst.

Fluid coking a continuous fluidized solid process that cracks feed thermally over heated coke particles in a reactor vessel to gas, liquid products, and coke.

Fouling an accumulation of deposits; this includes accumulation and growth of marine organisms on a submerged metal surface and the accumulation of deposits (usually inorganic) on heat exchanger tubing.

Fractional composition the composition of petroleum as determined by fractionation (separation) methods.

Fractional distillation the separation of the components of a liquid mixture by vaporizing and collecting the fractions, or cuts, which condense in different temperature ranges.

Fractionating column a column arranged to separate various fractions of petroleum by a single distillation and which may be tapped at different points along its length to separate various fractions in the order of their boiling points.

Fractionation the separation of petroleum into the constituent fractions using solvent or adsorbent methods; chemical agents such as sulfuric acid may also be used.

Fuel oil also called heating oil is a distillate product that covers a wide range of properties; see also No. 1 to No. 4 fuel oils.

Fuller's earth a clay which has high adsorptive capacity for removing color from oils; attapulgus clay is a widely used fuller's earth.

Functional group the portion of a molecule that is characteristic of a family of compounds and determines the properties of these compounds.

Furnace oil a distillate fuel primarily intended for use in domestic heating equipment.

Gaseous corrosion corrosion with gas as the only corrosive agent and without any aqueous phase on the surface of the metal, also called *dry corrosion*.

Gasoline fuel for the internal combustion engine that is commonly, but improperly, referred to simply as gas.

Gas oil a petroleum distillate with a viscosity and boiling range between those of kerosine and lubricating oil.

General corrosion corrosion in a uniform manner; corrosion that is distributed more or less uniformly over the surface of a material.

Graphitic corrosion deterioration of gray cast iron in which the metallic constituents are selectively leached or converted to corrosion products, leaving the graphite intact.

Gravimetric analysis a method of analysis whereby the dry weight of contaminant per unit volume of fluid can be measured showing the degree of contamination in terms of milligrams of contaminant per liter of fluid.

Gravity see Specific Gravity; API Gravity.

Gravity separation a method of separating two components from a mixture. Under the influence of gravity, separation of immiscible phases (gas–solid, liquid–solid, liquid–liquid, solid–solid) allows the denser phase to settle out.

Gray clay treating a fixed bed (*q.v.*), usually fuller's earth (*q.v.*), vapor phase treating process to selectively polymerize unsaturated gum-forming constituents (diolefins) in thermally cracked gasoline.

Gray iron a broad class of ferrous casting alloys (cast irons) normally characterized by a microstructure of flake graphite in a ferrous matrix; gray iron usually contains 2.5–4% (w/w) C, 1–3% (w/w) Si, and additions of manganese, depending on the desired microstructure (as low as 0.1% w/w Mn in ferritic gray iron); sulfur and phosphorus are also present in small amounts as residual impurities.

Guard bed a bed of an adsorbent (such as, e.g., bauxite) that protects a catalyst bed by adsorbing species detrimental to the catalyst.

Gum an insoluble tacky semisolid material formed as a result of the storage instability and/or the thermal instability of petroleum and petroleum products.

Hard vacuum a term used to denote a high vacuum.

HC hydrocarbon.

Heat affected zone (HAZ) an area adjacent to a weld where the thermal cycle has caused microstructural changes which generally affect corrosion behavior.

Heat exchanger a device which transfers heat through a conducting wall from one fluid to another.

Heating oil see Fuel oil.

Heavy ends the portions of a petroleum distillate fraction which are highest boiling, and therefore distill over last if the temperature is raised progressively.

Heavy fuel oil fuel oil having a high density and viscosity; generally residual fuel oil such as No. 5 and No 6. fuel oil (*q.v.*).

Heavy oil typically petroleum having an API gravity of $<20°$.

Heavy petroleum see Heavy oil.

Hematite (i) an iron mineral crystallizing in the rhombohedral system, the most important ore of iron and (ii) an iron oxide (Fe_2O_3) corresponding to an iron content of approximately 70% (w/w).

Heteroatom compounds chemical compounds which contain nitrogen and/or oxygen and/or sulfur and/or metals bound within their molecular structure(s).

Hot corrosion an accelerated corrosion of metal surfaces that results from the combined effect of oxidation and reactions with sulfur compounds and other contaminants, such as chlorides, to form a molten salt on a metal surface that fluxes, destroys, or disrupts the normal protective oxide.

Hydrocarbon an organic compound containing carbon and hydrogen *only*.

Hydrocarbon compounds (hydrocarbons) chemical compounds containing only carbon and hydrogen.

Hydroconversion a term often applied to hydrocracking (*q. v.*).

Hydrocracking a catalytic high-pressure high temperature process for the conversion of petroleum feedstocks in the presence of fresh and recycled hydrogen; carbon–carbon bonds are cleaved in addition to the removal of heteroatomic species; a process combining cracking or pyrolysis, with hydrogenation; feedstocks can include crude oil, distillates, heavy oil, tar sand bitumen, and residua.

Hydrocracking catalyst a catalyst used for hydrocracking which typically contains separate hydrogenation and cracking functions.

Hydrodenitrogenation the removal of nitrogen by hydrotreating (*q. v.*).

Hydrodesulfurization the removal of sulfur by hydrotreating (*q. v.*).

Hydrofinishing a process for treating raw extracted base stocks with hydrogen to saturate them for improved stability.

Hydrogenation the chemical addition of hydrogen to a hydrocarbon in the presence of a catalyst; a severe form of hydrogen treating. Hydrogenation may be either destructive or nondestructive.

Hydrogen blistering the formation of subsurface planar cavities, called hydrogen blisters, in a metal resulting from excessive internal hydrogen pressure; growth of near-surface blisters in low strength metals usually results in surface bulges.

Hydrogen damage a general term for the embrittlement, cracking, blistering, and hydride formation that can occur when hydrogen is present in some metals.

Hydrogen embrittlement a loss of ductility of a metal resulting from absorption of hydrogen; hydrogen-induced cracking (HIC) or severe loss of ductility caused by the presence of hydrogen in the metal; often generated from atomic hydrogen produced from cathodic reactions.

Hydrogen-induced cracking stepwise internal cracks that connect adjacent hydrogen blisters on different planes in the metal or to the metal surface (also known as *stepwise cracking*).

Hydrogen probes probes designed to measure the permeation rate of atomic hydrogen H + (measured as hydrogen gas, H_2) associated with hydrogen-induced cracking.

Hydrogen refining lube oil hydrorefining and hydrocracking or severe hydrotreating processes.

Hydrolysis breakdown process that occurs in anhydrous hydraulic fluids as a result of heat, water, and metal catalysts (iron, steel, copper, etc.).

Hydrophilic compounds with an affinity for water.

Hydrophobic compounds that repel water.

Hydroprocessing a term often equally applied to hydrotreating (*q.v.*) and to hydrocracking (*q.v.*); also often collectively applied to both.

Hydrotreating the removal of heteroatomic (nitrogen, oxygen, and sulfur) species by treatment of a feedstock or product at relatively low temperatures in the presence of hydrogen.

Hydrovisbreaking a noncatalytic process, conducted under similar conditions to visbreaking, which involves treatment with hydrogen to reduce the viscosity of the feedstock and produce more stable products than is possible with visbreaking.

IBP initial boiling point.

Immiscibility the inability of two or more fluids to have complete mutual solubility; they coexist as separate phases.

Immiscible incapable of being mixed without separation of phases. Water and petroleum oil are immiscible under most conditions, although they can be made miscible with the addition of an emulsifier.

Impingement corrosion a form of erosion corrosion generally associated with the local impingement of a high velocity, flowing fluid against a solid surface.

Incompatibility the *immiscibility* of petroleum products and also of different crude oils which is often reflected in the formation of a separate phase after mixing and/or storage.

Incompatible fluids fluids which when mixed in a system will have a deleterious effect on that system, its components, or its operation.

Incubation period a period prior to the detection of corrosion while the metal is in contact with a corrosive agent.

Infrared spectroscopy an analytical technique that quantifies the vibration (stretching and bending) that occurs when a molecule absorbs (heat) energy in the infrared region of the electromagnetic spectrum.

Inhibitor a chemical substance or combination of substances that, when present in the proper concentration and forms in the environment, prevents or reduces corrosion.

Inhibitor a substance, the presence of which, in small amounts, in a petroleum product prevents or retards undesirable chemical changes from taking place in the product, or in the condition of the equipment in which the product is used.

Initial boiling point the recorded temperature when the first drop of liquid falls from the end of the condenser.

Instability the inability of a petroleum product to exist for periods of time without change to the product.

Interface the thin surface area separating two immiscible fluids that are in contact with each other.

Interfacial film a thin layer of material at the interface between two fluids which differs in composition from the bulk fluids.

Interfacial tension the strength of the film separating two immiscible fluids, e.g., oil and water or microemulsion and oil; measured in dynes (force) per centimeter or millidynes per centimeter.

Interfacial viscosity the viscosity of the interfacial film between two immiscible liquids.

Iodine number a measure of the iodine absorption by oil under standard conditions, used to indicate the quantity of unsaturated compounds present, also called iodine value.

IP Institute of Petroleum (UK).

ISO International Standards Organization which sets viscosity reference scales.

Jet fuel fuel meeting the required properties for use in jet engines and aircraft turbine engines.

Kaolinite a clay mineral formed by hydrothermal activity at the time of rock formation or by chemical weathering of rock with high feldspar content; usually associated with intrusive granite rock with high feldspar content.

Kata-condensed aromatic compounds compounds based on linear condensed aromatic hydrocarbon systems, e.g., anthracene and naphthacene (tetracene).

Kerosene (kerosine) a fraction of petroleum that was initially sought as an illuminant in lamps; a precursor to diesel fuel.

K-factor see Characterization factor.

Kinematic viscosity the ratio of viscosity (*q.v.*) to density, both measured at the same temperature; the time required for a fixed amount of an oil to flow through a capillary tube under the force of gravity. The unit of kinematic viscosity is the stoke or centistoke (1/100 of a stoke). Kinematic viscosity may be defined as the quotient of the absolute viscosity in centipoises divided by the specific gravity of a fluid, both at the same temperature.

Lamellar corrosion see Exfoliation corrosion.

Light ends low boiling volatile materials in a petroleum fraction. They are often unwanted and undesirable, but in gasoline the proportion of light ends deliberately included is used to assist low temperature starting.

Light hydrocarbons hydrocarbons with molecular weights less than that of heptane (C_7H_{16}).

Light oil the products distilled or processed from crude oil up to, but not including, the first lubricating oil distillate.

Light petroleum petroleum having an API gravity $>20°$.

Liquid any substance that flows readily or changes in response to the smallest influence. More generally, any substance in which the force required to produce a deformation depends on the rate of deformation rather than on the magnitude of the deformation.

Liquid petrolatum see White oil.

Liquid chromatography a chromatographic technique that employs a liquid mobile phase.

Liquid/liquid extraction an extraction technique in which one liquid is shaken with or contacted by an extraction solvent to transfer molecules of interest into the solvent phase.

Localized corrosion the corrosion process in which an intense attack takes place only in and around particular zones of the metal, leaving the rest of the metal unaffected; an example is pitting corrosion.

Low carbon steel steel having $<0.30\%$ carbon and no intentional alloying additions.

Lubricating oil a fluid lubricant used to reduce friction between bearing surfaces.

Mahogany acids oil-soluble sulfonic acids formed by the action of sulfuric acid on petroleum distillates. They may be converted to their sodium soaps (mahogany soaps) and extracted from the oil with alcohol for use in the manufacture of soluble oils, rust preventives, and special greases. The calcium and barium soaps of these acids are used as detergent additives in motor oils; see also Brown acids and Sulfonic acids.

Maltenes that fraction of petroleum that is soluble in, e.g., pentane or heptane; deasphalted oil (*q.v.*); also the term arbitrarily assigned to the pentane-soluble portion of petroleum that is relatively high boiling (> 300°C, 760 mm) (see also Petrolenes).

Mass spectrometer an analytical technique that *fractures* organic compounds into characteristic "fragments" based on functional groups that have a specific mass-to-charge ratio.

Magnetite a naturally occurring magnetic oxide of iron (Fe_3O_4).

Material Safety Data Sheet (MSDS) a publication containing health and safety information on a hazardous product (including petroleum). The OSHA Hazard Communication Standard requires that an MSDS be provided by manufacturers to distributors or purchasers prior to or at the time of product shipment. An MSDS must include the chemical and common names of all ingredients that have been determined to be health hazards if they constitute 1% or greater of the product's composition (0.1% for carcinogens). An MSDS also included precautionary guidelines and emergency procedures.

mg milligram, unit of mass.

Microbial corrosion (bacterial corrosion) corrosion caused or promoted by microorganisms. It can apply to both metals and nonmetallic materials, in both the presence and lack of oxygen. Sulfate-reducing bacteria are common in anaerobic conditions (lack of oxygen) and produce hydrogen sulfide which can cause sulfide stress cracking. Under aerobic conditions (presence of oxygen), some bacteria directly oxidize iron to iron oxides and hydroxides which other bacteria oxidize sulfur and produce sulfuric acid causing biogenic sulfide corrosion. Concentration cells can form in the deposits of corrosion products, causing and enhancing galvanic corrosion; see Biocorrosion, Microbiologically influenced corrosion.

Microbiologically influenced corrosion (MIC) corrosion which is substantially increased as the result of the presence of bacteria, such as sulfate-reducing bacteria (SRB) or acid producing bacteria (APB).

Microcarbon residue the carbon residue determined using a themogravimetric method; see also Carbon residue.

Microemulsion a stable, finely dispersed mixture of oil, water, and chemicals (surfactants and alcohols).

Microemulsion or micellar/emulsion flooding an augmented waterflooding technique in which a surfactant system is injected in order to enhance oil displacement toward producing wells.

Microorganisms animals or plants of microscopic size, such as bacteria.

Mid-boiling point the temperature at which approximately 50% of a material has distilled under specific conditions.

Middle distillate distillate boiling between the kerosene and lubricating oil fractions.

Mill scale the oxide layer formed during hot fabrication or heat treatment of metals.

Mineral hydrocarbons petroleum hydrocarbons, considered *mineral* because they come from the earth rather than from plants or animals.

Mineral oil oil derived from a mineral source, such as petroleum, as opposed to oils derived from plants and animals; the older term for petroleum; the term was introduced in the nineteenth century as a means of differentiating petroleum (rock oil) from whale oil which, at the time, was the predominant illuminant for oil lamps.

Mineral seal oil a distillation fraction between kerosene and gas oil, widely used as a solvent oil in gas adsorption processes, as a lubricant for the rolling of metal foil, and as a base oil in many specialty formulations. Mineral seal oil takes its name—not from any sealing function—but from the fact that it originally replaced oil derived from seal blubber for use as an illuminant for signal lamps and lighthouses.

Miscibility an equilibrium condition, achieved after mixing two or more fluids, which is characterized by the absence of interfaces between the fluids. (i) *First-contact miscibility*: miscibility in the usual sense, whereby two fluids can be mixed in all proportions without any interfaces forming. Example: At room temperature and pressure, ethyl alcohol and water are first-contact miscible. (ii) *Multiple-contact miscibility (dynamic miscibility)*: miscibility that is developed by repeated enrichment of one fluid phase with components from a second fluid phase with which it comes into contact. (iii) *Minimum miscibility* pressure: the minimum pressure above which two fluids become miscible at a given temperature, or can become miscible, by dynamic processes.

Miscible capable of being mixed in any concentration without separation of phases; e.g., water and ethyl alcohol are miscible.

Mitigation identification, evaluation, and cessation of potential impacts of a process product or by-product.

MSDS Material Safety Data Sheet, a document required by several government agencies which typically lists the composition of a product along with hazard information, first aid measures, toxicological information, and regulatory information.

NACE (NACE International) National Association of Corrosion Engineers.

Naphtha a generic term applied to refined, partly refined, or unrefined petroleum products and liquid products of natural gas, the majority of which distills below 240°C (464°F); the volatile fraction of petroleum which is used as a solvent or as a precursor to gasoline.

Naphthenes a group of cyclic hydrocarbons also termed cycloparaffins; polycyclic members are also found in the higher boiling fractions.

Naphthenic a type of petroleum fluid derived from naphthenic crude oil, containing a high proportion of closed-ring methylene groups.

Naphthenic crude oil class designation of crude oil containing predominantly naphthenes or asphaltic compounds.

Native asphalt see Bitumen.

Natural asphalt see Bitumen.

Natural gas the naturally occurring gaseous constituents that are found in many petroleum reservoirs; there are also those reservoirs in which natural gas may be the sole occupant.

Natural gas liquids (NGL) the hydrocarbon liquids that condense during the processing of hydrocarbon gases that are produced from oil or gas reservoir; see also Natural gasoline.

Natural gasoline a mixture of liquid hydrocarbons extracted from natural gas (*q.v.*) suitable for blending with refinery gasoline.

Neutralization a process for reducing the acidity or alkalinity of a waste stream by mixing acids and bases to produce a neutral solution, also known as pH adjustment.

Neutralization number the weight, in milligrams, of potassium hydroxide needed to neutralize the acid in 1 g of oil; an indication of the acidity of crude oil or crude oil products; this number (as determined) includes organic or inorganic acids or bases or a combination thereof (ASTM D974).

Nonionic surfactant a surfactant molecule containing no ionic charge, a neutral surfactant.

Normal paraffin (n-paraffin) a hydrocarbon consisting of molecules in which any carbon atom is attached to no more than two other carbon atoms, also called straight chain paraffin and linear paraffin.

Nuclear magnetic resonance spectroscopy an analytical procedure that permits the identification of complex molecules based on the magnetic properties of the atoms they contain.

Number 1 Fuel oil (No. 1 Fuel oil) very similar to kerosene (*q.v.*) and is used in burners where vaporization before burning is usually required and a clean flame is specified.

Number 2 Fuel oil (No. 2 Fuel oil) also called domestic heating oil; has properties similar to diesel fuel and heavy jet fuel; used in burners where complete vaporization is not required before burning.

Number 4 Fuel oil (No. 4 Fuel oil) a light industrial heating oil and is used where preheating is not required for handling or burning; there are two grades of No. 4 fuel oil, differing in safety (flash point) and flow (viscosity) properties.

Number 5 Fuel oil (No. 5 Fuel oil) a heavy industrial fuel oil which requires preheating before burning.

Number 6 Fuel oil (No. 6 Fuel oil) a heavy fuel oil and is more commonly known as Bunker C oil when it is used to fuel oceangoing vessels; preheating is always required for burning this oil.

Oil the portion of petroleum that exists in the liquid phase in reservoirs and remains as such under original pressure and temperature conditions. Small amounts of nonhydrocarbon substances may be included.

Oil sand see Tar sand.

Oil shale a fine-grained impervious sedimentary rock which contains an organic material called kerogen.

Olefin synonymous with *alkene*.

Overhead that portion of the feedstock which is vaporized and removed during distillation.

Oxidation a process which can be used for the treatment of a variety of inorganic and organic substances; also: (i) loss of electrons by a constituent of a chemical reaction. (ii) Corrosion of a metal that is exposed to an oxidizing gas at elevated temperatures.

P Poise, unit of dynamic viscosity.

Paraffin any hydrocarbon identified by saturated straight (normal) or branched (iso) carbon chains, also called an alkane. The generalized paraffinic molecule can be symbolized by the formula C_nH_{2n+2}. Paraffins are relatively nonreactive and have excellent oxidation stability. In contrast to naphthenic oils, paraffinic lubricating oils have relatively high wax content and pour point, and generally have a high viscosity index (VI). Paraffinic solvents are generally lower in solvency than naphthenic or aromatic solvents.

Paraffinic a type of petroleum fluid derived from paraffinic crude oil and containing a high proportion of straight chain saturated hydrocarbons; often susceptible to cold flow problems.

Parting see Dealloying.

Patina a thin layer of corrosion product, usually green, that forms on the surface of metals such as copper and copper-based alloys exposed to the atmosphere.

Partitioning in chromatography, the physical act of a solute having different affinities for the stationary and mobile phases.

Partition ratios, K the ratio of total analytical concentration of a solute in the stationary phase, CS, to its concentration in the mobile phase, CM.

Pericondensed aromatic compounds compounds based on angular condensed aromatic hydrocarbon systems, e.g., phenanthrene, chrysene, and picene.

Petrochemical any chemical substance derived from crude oil or its products, or from natural gas. Some petrochemical products may be identical to others produced from other raw materials such as coal and producer gas.

Petrolenes the term applied to that part of the pentane-soluble or heptane-soluble material that is low boiling ($<300°C$, $<570°F$,

760 mm) and can be distilled without thermal decomposition (see also Maltenes).

Petroleum (crude oil) a naturally occurring mixture of gaseous, liquid, and solid hydrocarbon compounds usually found trapped deep underground beneath impermeable cap rock and above a lower dome of sedimentary rock such as shale; most petroleum reservoirs occur in sedimentary rocks of marine, deltaic, or estuarine origin.

Petroleum refinery see Refinery.

Petroleum refining a complex sequence of events that result in the production of a variety of products.

pH measure of alkalinity or acidity in water and water-containing fluids. pH can be used to determine the corrosion-inhibiting characteristic in water-based fluids. Typically, pH > 8.0 is required to inhibit corrosion of iron and ferrous alloys in water-based fluids.

pH adjustment neutralization by the addition of acid or alkali.

Phase a separate fluid that coexists with other fluids; gas, oil, water, and other stable fluids such as microemulsions are all called phases in EOR research.

Phase behavior the tendency of a fluid system to form phases as a result of changing temperature, pressure, or the bulk composition of the fluids or of individual fluid phases.

Phase separation the formation of a separate phase that is usually the prelude to coke formation during a thermal process; the formation of a separate phase as a result of the instability/incompatibility of petroleum and petroleum products.

Physical and chemical properties the results from several analytical tests that measure various physical characteristics and ingredients (constituents) of an engine oil.

Pipe still a still in which heat is applied to the oil while being pumped through a coil or pipe arranged in a suitable firebox.

Pipestill gas the most volatile fraction that contains most of the gases that are generally dissolved in the crude. Also known as pipestill light ends.

Pipestill light ends see Pipestill gas.

Pitch the nonvolatile, brown to black, semisolid to solid viscous product from the destructive distillation of many bituminous or other organic materials, especially coal.

Pitting highly localized corrosion resulting in deep penetration at only a few spots; a type of corrosion in which there is loss of metal in localized areas—the corrosion rate in the pits is many times greater than the corrosion rate on the entire surface—and the resultant pits can be large and shallow or narrow and deep. Pitting is a more dangerous problem than general corrosion because the pitted areas can be easily penetrated.

Pitting corrosion observed on passive metals in presence of certain anions (in particular chloride) when the potential exceeds a critical value. This process typically produces cavities with diameters on the order of several tens of micrometers.

Pitting factor the ratio of the depth of the deepest pit resulting from corrosion divided by the average penetration as calculated from mass loss.

PNA (polynuclear aromatic) any of numerous complex hydrocarbon compounds consisting of three or more benzene rings in a compact molecular arrangement. Some types of PNAs are formed in fossil fuel combustion and other heat processes such as catalytic cracking.

Poise (absolute viscosity) a measure of viscosity numerically equal to the force required to move a plane surface of one square centimeter per second when the surfaces are separated by a layer of fluid one centimeter in thickness.

Polar aromatics resins; the constituents of petroleum that are predominantly aromatic in character and contain polar (nitrogen, oxygen, and sulfur) functions in their molecular structure(s).

Polar compound a chemical compound whose molecules exhibit electrically positive characteristics at one extremity and negative characteristics at the other.

Polycyclic aromatic hydrocarbons (PAHs) polycyclic aromatic hydrocarbons are a suite of compounds comprised of two or more condensed aromatic rings. They are found in many petroleum mixtures, and they are predominantly introduced to the environment through natural and anthropogenic combustion processes.

Polynuclear aromatic compound an aromatic compound having two or more fused benzene rings, e.g., naphthalene and phenanthrene.

Pour point the lowest temperature at which oil will pour or flow when it is chilled (cooled) without disturbance under prescribed conditions (ASTM D97); the pour point is 3°C (5°F) above the temperature at which the oil in a test vessel shows no movement when the container is held horizontally for 5 s.

Pour point depressant an additive which retards the adverse effects of wax.

ppm parts per million (1/ppm = 0.000001); generally by weight (ppm w/w); 100 ppm = 0.01%; 10,000 ppm = 1%.

Precipitation number the number of milliliters of precipitate formed when 10 ml of lubricating oil is mixed with 90 ml of petroleum naphtha of a definite quality and centrifuged under definitely prescribed conditions.

Pressure, absolute the sum of atmospheric and gage pressures.

Pretreatment usually refers to the treatment of a crude oil or crude oil product prior to refining.

Primary treatment the first step in re-refining used lubricating using dehydration in which the oil is stored to allow water and solids to separate out from the oil, then the oil is heated to approximately 120°C (248°F) in a closed vessel to boil off any emulsified water and some of the fuel diluents.

Propane asphalt see Solvent asphalt.

Propane deasphalting solvent deasphalting using propane as the solvent.

Propane decarbonizing a solvent extraction process used to recover catalytic cracking feed from heavy fuel residues.

Propane dewaxing a process for dewaxing lubricating oils in which propane serves as solvent.

Propane fractionation a continuous extraction process employing liquid propane as the solvent; a variant of propane deasphalting (*q.v.*).

Protopetroleum a generic term used to indicate the initial product formed and changes have occurred to the precursors of petroleum.

psi pounds per square inch.

psia pounds per square inch absolute (psig + 14.696); absolute, unit of pressure (pressure above total vacuum).

psid pounds per square inch differential.

psig pounds per square inch gage (psia − 14.696); unit of pressure (pressure above atmospheric).

Quadrillion 1×10^{15}.

Quench the sudden cooling of hot material discharging from a thermal reactor.

Radiography (X-ray) use of X-rays to measure thickness or imperfection within solid materials.

Ramsbottom carbon residue an alternate test method for measuring the carbon residue of petroleum fractions (see Conradson carbon, Carbon residue).

Reaction a chemical transformation or change brought about by the interaction of two substances.

Reactive metal a metal that readily combines with oxygen at elevated temperatures to form very stable oxides, e.g., titanium, zirconium, and beryllium.

Reduced crude a residual product remaining after the removal, by distillation or other means, of an appreciable quantity of the more volatile components of crude oil.

Reducing agent a compound that causes reduction, thereby itself becoming oxidized.

Reducing atmosphere an atmosphere which tends to (i) promote the removal of oxygen from a chemical compound and (ii) promote the reduction of immersed materials.

Refinery a series of integrated unit processes by which petroleum can be converted to a slate of useful (salable) products.

Refinery gas a gas (or a gaseous mixture) produced as a result of refining operations.

Refining the processes by which petroleum is distilled and/or converted by application of a physical and chemical processes to form a variety of products are generated.

Relative humidity the ratio, expressed as a percentage, of the amount of water vapor present in a given volume of air at a given temperature to the amount required to saturate the air at that temperature.

Rerunning the distillation of an oil which has already been distilled.

Residuum (resid; *pl:*. residua) the residue obtained from petroleum after nondestructive distillation has removed all the volatile materials from crude oil, e.g., an atmospheric (345°C, 650°F +) residuum; the highest boiling (usually the highest molecular weight) components or bottoms remaining from distilling an oil, especially crude oil.

Resins that portion of the maltenes (*q.v.*) that is adsorbed by a surface active material such as clay or alumina; the fraction of deasphaltened oil that is insoluble in liquid propane but soluble in *n*-heptane.

Ringworm corrosion localized corrosion frequently observed in oil well tubing in which a circumferential attack is observed near a region of metal "upset."

Riser (i) the section of pipeline extending from the ocean floor up the platform, (ii) the vertical tube in a steam generator convection bank that circulates water and steam upward; the part of the bubble plate assembly which channels the vapor and causes it to flow downward to escape through the liquid; also the vertical pipe where fluid catalytic cracking reactions occur.

ROSE Process Residual Oil Supercritical Extraction Process (Kellogg Brown and Root).

Rust a corrosion product consisting primarily of hydrated iron oxide; a term properly applied only to ferrous alloys.

Rust bloom discoloration indicating the beginning of rusting.

Rusting the formation of an oxide of iron due to oxidation of the iron atoms in solid solution is a well-known example of electrochemical corrosion.

Salinity the concentration of salt in water.

Saponification number the number of milligrams of potassium hydroxide (KOH) that combine with one gram of oil under specified conditions (ASTM D94); an indication of the amount of fatty saponifiable material in compounded oil. Caution must be used in interpreting test results if certain substances—such as sulfur compounds or halogens—are present in the oil, since these also react with KOH thereby increasing the apparent Saponification number.

SARA analysis a method of fractionation by which petroleum is separated into saturates, aromatics, resins, and asphaltene fractions.

SARA separation see SARA analysis.

Saturates paraffins and cycloparaffins (naphthenes).

Saybolt Furol viscosity the time, in seconds (Saybolt Furol Seconds, SFS), for 60 ml of fluid to flow through a capillary tube in a Saybolt

Furol viscometer at specified temperatures between 70 and 210°F; the method is appropriate for high viscosity oils such as transmission, gear, and heavy fuel oils (ASTM D88).

Saybolt Universal viscosity the time, in seconds (Saybolt Universal Seconds, SUS), for 60 ml of fluid to flow through a capillary tube in a Saybolt Universal viscometer at a given temperature (ASTM D88).

Scaling (i) the formation at high temperatures of thick corrosion product layers on a metal surface; (ii) the deposition of water-insoluble constituents on a metal surface.

Sediment an insoluble solid formed as a result of the storage instability and/or the thermal instability of petroleum and petroleum products.

Selective corrosion (selective leaching or dealloying) the selective dissolution of one of the components of an alloy that forms a solid solution. It leads to the formation of a porous layer made of the more noble metal; the selective corrosion of certain alloying constituents from an alloy (as dezincification) or in an alloy (as internal oxidation).

Selective solvent a solvent which, at certain temperatures and ratios, will preferentially dissolve more of one component of a mixture than of another and thereby permit partial separation.

Separation process an upgrading process in which the constituents of petroleum are separated, usually without thermal decomposition, e.g., distillation and deasphalting.

Siderite an iron ore ($FeCO_3$) occurring in various forms and colors and crystalline with perfect rhombohedral cleavage.

Sidestream a liquid stream taken from any one of the intermediate plates of a bubble tower.

Sidestream stripper a device used to perform further distillation on a liquid stream from any one of the plates of a bubble tower, usually by the use of steam.

Sludge insoluble material formed as a result either of deterioration reactions in crude oil (crude oil product) or of contamination of the oil, or both; a semisolid to solid product which results from the storage instability and/or the thermal instability of petroleum and petroleum products.

Solvency ability of a fluid to dissolve inorganic materials and polymers, which is a function of aromaticity.

Solvent a material with a strong capability to dissolve a given substance. The most common petroleum solvents are mineral spirits, xylene, toluene, hexane, heptane, and naphtha. Aromatic-type solvents have the highest solvency for organic chemical materials, followed by naphthenes and paraffins. In most applications, the solvent disappears, usually by evaporation, after it has served its purpose. The evaporation rate of a solvent is very important in manufacture.

Solvent asphalt the asphalt (*q.v.*) produced by solvent extraction of residua (*q.v.*) or by light hydrocarbon (propane) treatment of a residuum (*q.v.*) or an asphaltic crude oil.

Solvent deasphalting a process for removing asphaltic and resinous materials from reduced crude oils, lubricating oil stocks, gas oils, or middle distillates through the extraction or precipitant action of low molecular weight hydrocarbon solvents; see also Propane deasphalting.

Solvent decarbonizing see Propane decarbonizing.

Solvent deresining see Solvent deasphalting.

Solvent dewaxing a process for removing wax from oils by means of solvents usually by chilling a mixture of solvent and waxy oil, filtration or by centrifuging the wax which precipitates, and solvent recovery.

Solvent extraction a process for separating liquids by mixing the stream with a solvent that is immiscible with part of the waste but that will extract certain components of the waste stream; a refining process used to separate components (unsaturated hydrocarbons) from lube distillates in order to improve the oil's oxidation stability, viscosity index, and response to additives.

Solvent naphtha a refined naphtha of restricted boiling range used as a solvent, also called petroleum naphtha, petroleum spirits.

Solvent refining see Solvent extraction.

Sour crude oil crude oil containing an abnormally large amount of sulfur compounds; see also Sweet crude oil.

Sour gas a gaseous mixture containing hydrogen sulfide and carbon dioxide.

Sour water waste water containing fetid materials, usually sulfur compounds.

Spalling the spontaneous chipping, fragmentation, or separation of a surface or surface coating.

Specific gravity the ratio of the weight of a given volume of material to the weight of an equal volume of water.

Specific gravity (liquid) the ratio of the weight of a given volume of liquid to the weight of an equal volume of water.

Spent catalyst catalyst that has lost much of its activity due to the deposition of coke and metals.

SSU Saybolt Universal Seconds (or SUS), a unit of measure used to indicate kinematic viscosity, e.g., SSU @ 100°F.

St Stokes, unit of kinematic viscosity.

Stabilization the removal of volatile constituents from a higher boiling fraction or product (*q.v.* stripping); the production of a product which, to all intents and purposes, does not undergo any further reaction when exposed to the air.

Stoke (St) kinematic measurement of a fluid's resistance to flow defined by the ratio of the fluid's dynamic viscosity to its density.

Straight mineral oil petroleum containing no additives.

Straight oil mineral oil containing no additives.

Straight-run asphalt the asphalt (*q.v.*) produced by the distillation of asphaltic crude oil.

Straight-run products obtained from a distillation unit and used without further treatment.

Stress corrosion (stress-accelerated corrosion) corrosion which is accelerated by stress.

Stress corrosion cracking results from the combined action of corrosion and of mechanical stress; manifested by crack formation at stress levels well below the ultimate tensile strength of a material.

Sulfidation the reaction of a metal or alloy with a sulfur-containing species to produce a sulfur compound that forms on or beneath the surface on the metal or alloy.

Sulfide stress cracking cracking of a metal under the combined action of tensile stress and corrosion in the presence of water and hydrogen sulfide (a form of hydrogen stress cracking).

Sulfonic acids acids obtained by petroleum or a petroleum product with strong sulfuric acid.

Sulfur a common natural constituent of petroleum and petroleum products; limitations of the amount of sulfur are specified in the quality control of fuels, solvents, and other products.

Surface active material a chemical compound, molecule, or aggregate of molecules with physical properties that cause it to adsorb at the interface between *two* immiscible liquids, resulting in a reduction of interfacial tension or the formation of a microemulsion.

Surface tension the contractile surface force of a liquid by which it tends to assume a spherical form and to present the least possible surface. It is expressed in dynes/cm or ergs/cm^2.

Surfactant surface active agent that reduces interfacial tension of a liquid; used in petroleum may increase the oil's affinity for metals and other materials; a chemical, characterized as one that reduces interfacial resistance to mixing between oil and water or changes the degree to which water wets reservoir rock.

SUS (SSU): Saybolt Universal Seconds a measure of lubricating oil viscosity in the oil industry. The measuring apparatus is filled with a specific quantity of oil or other fluid and its flow time through a standardized orifice is measured in seconds. Fast flowing fluids (low viscosity) will have a low value; slow flowing fluids (high viscosity) will have a high value.

Sweet crude oil crude oil containing little sulfur; see also Sour crude oil.

Synthetic crude oil (syncrude) a hydrocarbon product produced by the conversion of coal, oil shale, or tar sand bitumen that resembles conventional crude oil; can be refined in a petroleum refinery (*q.v.*); the product of a thermal reaction rather than by extraction or refining.

TAN total acid number.

Tar the volatile, brown to black, oily, viscous product from the destructive distillation of many bituminous or other organic materials, especially coal; a name used for petroleum in ancient texts.

Tar sand see Bituminous sand.

TBN total base number.

Thermal coke the carbonaceous residue formed as a result of a non-catalytic thermal process, the Conradson carbon residue, the Ramsbottom carbon residue.

Thermal cracking a process which decomposes, rearranges, or combines hydrocarbon molecules by the application of heat, without the aid of catalysts.

Thermal process any refining process which utilizes heat, without the aid of a catalyst.

Thermal stability (thermal instability) the ability (inability) of a liquid to withstand relatively high temperatures for short periods of time without the formation of carbonaceous deposits (sediment or coke); the ability of crude oil or a crude oil product to resist degradation or oxidation under high temperature operating conditions.

Thin layer chromatography (TLC) a chromatographic technique employing a porous medium of glass coated with a stationary phase. An extract is spotted near the bottom of the medium and placed in a chamber with solvent (mobile phase). The solvent moves up the medium and separates the components of the extract, based on affinities for the medium and solvent.

Ton a short ton is 2000 lbs (907.2 kg); a long ton is 2240 lbs (1016 kg).

Tonne A metric tonne of 1000 kg, equivalent to 2205 lbs.

Topped crude petroleum that has had volatile constituents removed up to a certain temperature, e.g., 250°C+ (480°F+) topped crude; not always the same as a residuum (*q.v.*).

Topping the distillation of crude oil to remove light fractions only.

Total acid number (TAN) the quantity of base, expressed in milligrams of potassium hydroxide, that is required to neutralize all acidic constituents present in 1 gram of sample (ASTM D974); see Acid Number.

Total base number (TBN) the quantity of acid, expressed in terms of the equivalent number of milligrams of potassium hydroxide that is required to neutralize all basic constituents present in 1 gram of sample (ASTM D974); see Base Number.

Tower equipment for increasing the degree of separation obtained during the distillation of oil in a still.

Treatment any method, technique, or process that changes the physical and/or chemical character of petroleum.

Trillion 1×10^{12}.

True boiling point (True boiling range) the boiling point (boiling range) of a crude oil fraction or a crude oil product under standard conditions of temperature and pressure.

Tuberculation the formation of localized corrosion products scattered over the surface in the form of knob-like mounds called tubercles.

UK United Kingdom of Great Britain.

Ultimate analysis elemental composition.

Ultrasonic measurement The timing of the transmission of ultrasonic sound waves through a material to determine the material's thickness.

Underfilm corrosion see Filiform corrosion.

Uniform corrosion the loss of material distributed uniformly over the entire surface exposed to the corrosive environment; metals in contact with strong acids are sometimes subject to uniform corrosion.

Unstable usually refers to a petroleum product that has more volatile constituents present or refers to the presence of olefin and other unsaturated constituents.

Upgrading the conversion of petroleum to value-added salable products.

US or USA United States of America.

Vacuum dehydration a method which involves drying or freeing of moisture through a vacuum process.

Vacuum distillation a distillation method which involved reducing the pressure above a liquid mixture to be distilled to less than its vapor pressure (usually less than atmospheric pressure); this causes evaporation of the most volatile liquid(s)—those with the lowest boiling points; distillation (*q.v.*) under reduced pressure.

Vacuum residuum a residuum (*q.v.*) obtained by distillation of a crude oil under vacuum (reduced pressure), that portion of petroleum which boils above a selected temperature such as 510°C (950°F) or 565°C (1050°F).

Vacuum separator a separator that utilizes subatmospheric pressure to remove certain gases and liquids from another liquid because of their difference in vapor pressure.

Vapor pressure pressure of a confined vapor in equilibrium with its liquid at specified temperature, thus a measure of a liquid's volatility.

Vapor pressure-Reid (RVP) a measure of the pressure of vapor accumulated above a sample of gasoline or other volatile fuel in a standard bomb at 100°F (37.8°C).

VD vacuum distillation.

VDT (vacuum distillation tower) generally applies to a crude distillation tower which operates at below atmospheric pressure.

VDU (vacuum distillation unit) generally includes a VDT and associated equipment for producing distillates from the bottoms of an atmospheric distillation tower (ADT) by operating at below atmospheric pressure.

VI viscosity index.

Visbreaking a process for reducing the viscosity of heavy feedstocks by controlled thermal decomposition.

Viscometer (viscosimeter) an apparatus for determining the viscosity of a fluid.

Viscosity a measure of the ability of a liquid to flow or a measure of its resistance to flow; the force required to move a plane surface of area 1 m^2 over another parallel plane surface 1 m away at a rate of 1 m/s when both surfaces are immersed in the fluid; measurement of a fluid's resistance to flow; the common metric unit of absolute viscosity is the poise, which is defined as the force in dynes required to move a surface one square centimeter in area past a parallel surface at a speed of one centimeter per second, with the surfaces separated by a fluid film 1 cm thick. In addition to kinematic viscosity, there are other methods for determining viscosity, including Saybolt Universal Viscosity (SUV), Saybolt Furol viscosity, Engler viscosity, and Redwood viscosity. Since viscosity varies in inversely with temperature, its value is meaningless until the temperature at which the viscosity is determined is also reported.

VGC (viscosity-gravity constant) an index of the chemical composition of crude oil defined by the general relation between specific gravity, sg, at 60°F and Saybolt Universal viscosity, SUV, at 100°F:

$$a = 10\text{sg} - 1.0752 \log (\text{SUV} - 38)$$
$$/10\text{sg} - \log (\text{SUV} - 38)$$

The constant, a, is low for the paraffin crude oils and high for the naphthenic crude oils.

Viscosity-gravity constant see VGC.

Viscosity index (VI) a commonly used measure of a fluid's change of viscosity with temperature. The higher the viscosity index, the smaller the relative change in viscosity with temperature.

Viscosity–temperature relationship the manner in which the viscosity of a given fluid varies inversely with temperature.

Viscous frequently used to imply high viscosity.

Void (i) a holiday, hole, or skip in a coating, (ii) a hole in a casting or weld deposit usually resulting from shrinkage during cooling.

Volatile compounds a relative term that may mean (i) any compound that will purge, (ii) any compound that will elute before the solvent peak (usually those <C6), or (iii) any compound that will not evaporate during a solvent removal step.

Volatility this property describes the degree and rate at which a liquid will vaporize under given conditions of temperature and pressure. When liquid stability changes, this property is often reduced in value.

VPS (vacuum pipe still) generally includes a vacuum tower and associated equipment for the distillation of crude into lube distillates or cracking feedstocks and vacuum residua; see VDU.

Watson characterization factor see Characterization factor.

White oil a generic name applied to highly refined, colorless hydrocarbon oils of low volatility, and covering a wide range of viscosity; a colorless and odorless mineral oil used in medicinal and pharmaceutical preparations and as a lubricant in food and textile industries.

White rust zinc oxide (ZnO); the powdery product of the corrosion of zinc or zinc-coated surfaces.

Zeolite a crystalline aluminosilicate used as a catalyst and having a particular chemical and physical structure.

eBIBLIOGRAPHY

CHAPTER 1

Aitken, C.M., Jones, D.M., Larter, S.R., 2004. Anaerobic hydrocarbon biodegradation in deep subsurface oil reservoirs. Nature 431 (7006), 291–294.

ASTM D1534, 2013. Standard Test Method for Approximate Acidity in Electrical Insulating Liquids by Color-Indicator Titration. Annual Book of Standards, ASTM International, West Conshohocken, PA.

ASTM D2896, 2013. Standard Test Method for Base Number of Petroleum Products by Potentiometric Perchloric Acid Titration. Annual Book of Standards, ASTM International, West Conshohocken, PA.

ASTM D3339, 2013. Standard Test Method for Acid Number of Petroleum Products by Semi-Micro Color Indicator Titration. Annual Book of Standards, ASTM International, West Conshohocken, PA.

ASTM D664, 2013. Standard Test Method for the Acid Number of Petroleum Products by Potentiometric Titration. Annual Book of Standards, ASTM International, West Conshohocken, PA.

ASTM D974, 2013. Standard Test Method for Acid and Base Number by Color-Indicator Titration. Annual Book of Standards, ASTM International, West Conshohocken, PA.

Babaian-Kibala, E., Craig, H.L., Rusk, G.L., Quinter, R.C., Summers, M.A., 1993. Naphthenic acid corrosion in refinery settings. Mater. Perform. 32 (4), 50–55.

Babaian-Kibala, E., Petersen, P.R., Humphries, M.J., 1998. Corrosion by naphthenic acids in crude oils. Preprints. Div. Petrol. Chem., Am. Chem. Soc. 3, 106.

Baker Hughes., 2010. Planning Ahead for Effective Canadian Crude Processing. White Paper. Baker Petrolite, Sugar Land, TX.

Barrow, M.P., McDonnell, L.A., Feng, X., Walker, J., Derrick, P.J., 2003. Determination of the nature of naphthenic acids present in crude oils using nanospray Fourier transform ion cyclotron resonance mass spectrometry: the continued battle against corrosion. Anal. Chem. 75 (4), 860–866.

Barth, T., Høiland, S., Fotland, P., Askvik, K.M., Pedersen, B.S., Borgund, A.E., 2004. Acidic compounds in biodegraded petroleum. Org. Geochem. 35 (11–12), 1513–1525.

Bartha, R., Atlas, R.M., 1977. The microbiology of aquatic oil spills. Appl. Microbiol. 22, 225–266.

Behar, F.H., Albrecht, P., 1984. Correlations between carboxylic acids and hydrocarbons in several crude oils. Org. Geochem. 6, 597–604.

Bennett, B., Buckman, J.O., Bowler, B.F.J., Larter, S.R., 2004. Wettability alteration in petroleum systems: the role of polar non-hydrocarbons. Petrol. Geosci. 10 (3), 271–277.

Blume, A.M., Yeung, T.Y., 2008. Analyzing economic viability of opportunity crudes. Petrol. Technol. Q. Q3, 67–73.

Brient, J.A., Wessner, P.J., Doyle, M.N., 1995. Naphthenic acids. In: fourth ed. Howe-Grant, M. (Ed.), Encyclopedia of Chemical Technology, vol. 16. John Wiley & Sons, Inc., New York, NY, pp. 1017–1029.

Cai, X., Tian, S., 2011. Review and comprehensive analysis of composition and origin of high acidity crude oils. China Petrol. Process. Petrochem. Technol. 13 (1), 6–15.

Chevron, 2012a. Qin Huang Dao Crude Oil Assay. Available from: <http://crudemarketing.chevron.com/crude/far_eastern/qinghuangdao.aspx>.

Chevron, 2012b. Duri Crude Oil Assay. Available from: <http://crudemarketing.chevron.com/crude/far_eastern/duri.aspx>.

Clemente, J.S., Fedorak, P.M., 2005. A review of the occurrence, analyses, toxicity, and biodegradation of naphthenic acids. Chemosphere 60, 585–600.

Clemente, J.S., Yen, T.W., Fedorak, P.M., 2003a. Development of a high performance liquid chromatography method to monitor the biodegradation of naphthenic acids. J. Environ. Eng. Sci. 2 (3), 177–186.

Clemente, J.S., Prasad, N.G.N., MacKinnon, M.D., Fedorak, P.M., 2003b. A statistical comparison of naphthenic acids characterized by gas chromatography–mass spectrometry. Chemosphere 50, 1265–1274.

Clemente, J.S., MacKinnon, M.D., Fedorak, P.M., 2004. Aerobic biodegradation of two commercial naphthenic acids preparations. Environ Sci. Technol. 38, 1009–1016.

Conan, J., 1984. Biodegradation of crude oils in reservoirs. In: Brooks, J., Welte, D. (Eds.), Advances in Petroleum Geochemistry. Academic Press, London, UK, pp. 299–335.

Costantinides, G., Arich, G., 1967. Non-hydrocarbon compounds in petroleum. In: Nagy, B., Colombo, U. (Eds.), Fundamental Aspects of Petroleum Geochemistry. Elsevier Publishing Company, London, UK, pp. 109–175.

Craig, H.L., 1995. Naphthenic acid corrosion in the refinery. Paper No. 333. In: Proceedings of the CORROSION'95. NACE International, Houston, TX.

Craig, H.L., 1996. Temperature and velocity effects in naphthenic acid corrosion. Paper No. 603. In: Proceedings of the CORROSION'96. NACE International, Houston, TX.

Damasceno, F.C., Gruber, L.D.A., Geller, A.M., Vaz de Campos, M.C., Gomes, A.O., Guimarães, R.C.L., et al., 2014. Characterization of naphthenic acids using mass spectroscopy and chromatographic techniques: study of technical mixtures. Anal. Methods 6, 807–816.

Derungs, W.A., 1956. Naphthenic acid corrosion—an old enemy of the petroleum industry. Corrosion 12 (12), 617–622.

Dou, L., Cheng, D., Li, Z., Zhang, Z., Wang, J., 2013. Petroleum geology of the Fula sub-basin, Muglad Basin, Sudan. J. Petrol. Geol. 36 (1), 43–59.

Dzidic, I., Somerville, A.C., Raia, J.C., Hart, H.V., 1988. Determination of Naphthenic acids in California crudes and refinery wastewaters by fluoride ion chemical ionization mass spectrometry. Anal. Chem. 60 (13), 1318–1323.

Dzidic, I. < http://pubs.acs.org/action/doSearch?action=search&author=Somerville%2C+A.+C.&qsSearchArea=author>.

ESMAP, 2005. Crude Oil Price Differentials and Differences in Oil Qualities: A Statistical Analysis. ESMAP Technical Paper No. 81. Energy Sector Management Assistance Program, Washington, DC.

Erfan, M., 2011. A review of catalytic removal of chlorides from refinery streams and a critique of current analytical techniques for estimating chloride content. Petrol. Technol. Q. Q1, 1–10.

Ese, M.H., Kilpatrick, P.K., 2004. Stabilization of water-in-oil emulsions by naphthenic acids and their salts: model compounds, role of ph, and soap: acid ratio. J. Dispers. Sci. Technol. 25 (3), 253–261.

Fafet, A., Kergall, F., Da Silva, M., Behar, F., 2008. Characterization of acidic compounds in biodegraded oils. Org. Geochem. 39 (8), 1235–1242.

Fan, T.P., 1991. Characterization of naphthenic acids in petroleum by fast atom bombardment mass spectrometry. Energy Fuels 5 (3), 371–375.

Frank, R.A., Kavanagh, R., Burnison, B.K., Arsenault, G., Headley, J.V., Peru, K.M., et al., 2008. Toxicity assessment of collected fractions from an extracted naphthenic acid mixture. Chemosphere 72, 1309–1314.

Frank, R.A., Fischer, K., Kavanagh, R., Burnison, B.K., Arsenault, G., Headley, J.V., et al., 2009. Effect of carboxylic acid content on the acute toxicity of oil sands naphthenic acids. Environ Sci. Technol. 43 (2), 266–271.

Gary, J.G., Handwerk, G.E., Kaiser, M.J., 2007. Petroleum Refining: Technology and Economics, fifth ed. CRC Press, Taylor & Francis Group, Boca Raton, FL.

Gaylarde, C.C., Bento, F.M., Kelley, J., 1999. Microbial contamination of stored hydrocarbon fuels and its control. Rev. Microbiol. 30 (1), 1–10.

Ghoshal, S., Sainik, V., 2013. Monitor and minimize corrosion in High-TAN crude processing. Hydrocarbon Process. 92 (3), 35–38.

Grishchenkov, V.G., Townsend, R.T., McDonald, T.J., Autenrieth, R.L., Bonner, J.S., Boronin, A.M., 2000. Degradation of petroleum hydrocarbons by facultative anaerobic bacteria under aerobic and anaerobic conditions. Process. Biochem. 35 (9), 889–896.

Gruber, L.D.A., Damasceno, F.C., Caramão, E.B., Jacques, R.A., Geller, A.M., Vaz de Campos, M.C., 2012. Naphthenic acids in petroleum. Quím. Nova 35 (7), 1423–1433.

Handa, S.K., 2012. Heavy crude processing. Proceedings of the International Conference on Refining Challenges and the Way Forward. Indian Institute of Petroleum, New Delhi, India, April 16–17.

Hart, H.V., 1988. Determination of naphthenic acids in california crudes and refinery waste-waters by fluoride ion chemical ionization mass spectrometry. Anal. Chem. 60 (13), 1318–1323.

Hau, J.L., Mirabal, E.J., 1996. Experience with processing high sulfur naphthenic acid containing heavy crude oils. Paper No. LA96037. In: Proceedings of the Second NACE International, Latin American Region Corrosion Congress. NACE International, Houston, TX.

Hau, J.L., Yépez, O.J., Specht, M.I., Lorenzo, R., 1999. The iron powder test for naphthenic acid corrosion studies. Paper No. 379. In: Proceedings of the CORROSION'99. NACE International, Houston, TX.

Hau, J.X., Yépez, O.J., Torres, L.H., Vera, J.R., 2003. Measuring naphthenic acid corrosion potential with the Fe powder test. Rev. Metal. Madrid Vol. Extr., 116–123.

Havre, T., 2002. Formation of Calcium Naphthenate in Water/Oil Systems, Naphthenic Acid Chemistry and Emulsion Stability (Ph.D. thesis). Department of Chemical Engineering, Norwegian University of Science and Technology, Trondheim, Norway, October 2002.

Haynes, D., 2006. Naphthenic acid bearing refinery feedstocks and corrosion abatement. In: Proceedings of the AIChE Chicago Symposium, October.

Head, I.M., Jones, D.M., Larter, S.R., 2003. Biological activity in the deep subsurface and the origin of heavy oil. Nature 426 (6964), 344–352.

Headley, J.V., McMartin, D.W., 2004. A review of the occurrence and fate of naphthenic acids in aquatic environments. J. Environ. Sci. Health A Tox. Hazard. Subst. Environ. Eng. A39 (8), 1989–2010.

Headley, J.V., Peru, K.M., McMartin, D.W., Winkler, M., 2002. Determination of dissolved naphthenic acids in natural waters using negative-ion electrospray mass spectrometry. J. AOAC Int. 85, 182–187.

Hell, C.C., Medinger, E., 1874. Ueber das Vorkommen und die Zusammensetzung von Säuren im Rohpetroleum. Ber. Dtsch. Chem. Ges. 7 (2), 1216–1223.

Heller, J.J., Merick, R.D., Marquand, E.B., 1963. Corrosion of refinery equipment by naphthenic acid. Mater. Prot. 2 (9), 44.

Herman, D., Fedorak, P., Costerton, J., 1993. Biodegradation of cycloalkane carboxylic acids in oil sands tailings. Can. J. Microbiol. 39, 576–580.

Herman, D., Fedorak, P., MacKinnon, M., Costerton, J., 1994. Biodegradation of naphthenic acids by microbial populations indigenous to oil sands tailings. Can. J. Microbiol. 40, 467–477.

Hoeiland, S., Barth, T., Blokhus, A.M., Skauge, A., 2001. The effect of crude oil acid fractions on wettability as studied by interfacial tension and contact angles. J. Petrol. Sci. Eng. 30 (2), 91–103.

Hsu, C.S., Dechert, G.J., Robbins, W.K., Fukuda, E.K., 2000. Naphthenic acids in crude oils characterized by mass spectrometry. Energy Fuels 14, 217–223.

Hsu, C.S., Robinson, P.R. (Eds.), 2006. Practical Advances in Petroleum Processing Volume 1 and Volume 2. Springer Science, New York, NY.

Huang, H., Larter, S.R., Bowler, B.F.J., Oldenburg, T.B.P., 2004. A dynamic biodegradation model suggested by petroleum compositional gradients within reservoir columns from The Liaohe Basin, NE China. Org. Geochem. 35 (3), 299–316.

Hughey, C.A., Rodgers, R.P., Marshall, A.G., Qian, K., Robbins, W.K., 2002. Identification of acidic NSO compounds in crude oils of different geochemical origins by negative ion electrospray Fourier transform ion cyclotron resonance mass spectrometry. Org. Geochem. 33 (7), 743–759.

Hughey, C.A., Galasso, S.A., Zumberge, J.E., 2007. Detailed compositional comparison of acidic NSO compounds in biodegraded reservoir and surface crude oils by negative ion electrospray Fourier transform ion cyclotron resonance mass spectrometry. Fuel 86 (5-6), 758–768.

Johnson, D., McAteer, G., Zuk, H., 2003. The safe processing of high naphthenic acid content crude oils refinery experience and mitigations studies. Paper No. 03645. In: Proceedings of the CORROSION'03. NACE International, Houston, TX.

Jones, D.M., Watson, J.S., Meredith, W., Chen, M., Bennett, B., 2001. Determination of naphthenic acids in crude oils using nonaqueous ion exchange solid-phase extraction. Anal. Chem. 73, 703–707.

Juan, S.L., Xian, S.B., 2009. Separation and characterization of naphthenic acids contained in Penglai crude oil. Petrol. Sci. Technol. 27 (14), 1534–1544.

Kane, R.D., Cayard, M.S., 2002. A comprehensive study on naphthenic acid corrosion. In: CORROSION'02. NACE International, Houston, TX.

Kim, K., Stanford, L.A., Rodgers, R.P., Marshall, A.G., Walters, C.C., Qian, K., et al., 2005. Microbial alteration of the acidic and neutral polar NSO compounds revealed by Fourier transform ion cyclotron resonance mass spectrometry. Org. Geochem. 36 (8), 1117–1134.

Lai, J.W.S., Pinto, L.J., Kiehlmann, E., Bendell-Young, L.I., Moore, M.M., 1996. Factors that affect the degradation of naphthenic acids in oil sands wastewater by indigenous microbial communities. Environ. Toxicol. Chem. 15, 1482–1491.

Langevin, D., Poteau, S., Hénaut, I., Argillier, J.F., 2004. Crude oil emulsion properties and their application to heavy oil transportation. Oil Gas Sci. Technol., Rev. I. FR. Petrol. 59 (5), 511–521.

Laredo, G.C., Lopez, C.R., Alvarez, R.E., Cano, J.L., 2004. Naphthenic acids, total acid number and sulfur content profile characterization in isthmus and maya crude oils. Fuel 83, 1689–1695.

Lewis, K.R., Daane, M.L., Schelling, R., 1999. Processing corrosive crude oils. In: CORROSION'99. NACE International, Houston, TX.

Lochte, H.L., 1952. Petroleum acids and bases. Ind. Eng. Chem. 44 (11), 2597–2601.

MacKinnon, M.D., Boerger, H., 1986. Description of two treatment methods for detoxifying oil sands tailings pond water. Water Pollut. Res. J. Can. 21, 496–512.

Mackenzie, A.S., Wolff, G.A., Maxwell, J.R., 1981. Fatty acids in some biodegraded petroleums. Possible origins and significance. In: Bjorøy, M. (Ed.), Advances in Organic Geochemistry. John Wiley & Sons, Inc., Chichester, UK, pp. 637–649.

Magot, M., Ollivier, B., Patel, B.K.C., 2000. Microbiology of petroleum reservoirs. Antoine van Leeuwenhoek Int. J. Gen. Mol. Microbiol. 77 (2), 103–116.

Marshall, A.G., Rodgers, R.P., 2008. Petroleomics: chemistry of the underworld. Proc. Natl. Acad. Sci. USA 105 (47), 18090–18095.

McKenna, E.J., Kallio, R.E., 1965. The biology of hydrocarbons. Annu. Rev. Microbiol. 19, 183–208.

Meredith, W., Kelland, S.J., Jones, D.M., 2000. Influence of biodegradation on crude oil activity and carboxylic acid composition. Org. Geochem. 31 (11), 1059–1073.

Messer, B., Tarleton, B., Beaton, M., Phillips, T., 2004. New theory for naphthenic acid corrosivity of Athabasca oil sands crudes. Paper No. 04634. In: Proceedings of the CORROSION'04. NACE International, Houston, TX.

Moreira, A.P.D., Teixeira, A.M.R.F., 2009. An investigation on the formation of calcium naphthenate from commercial naphthenic acid solutions by thermogravimetric analysis. Braz. J. Petrol. Gas 3 (2), 51–56.

Ney, W.O., Crouch, W.W., Rannefeld, C.E., Lochte, H.L., 1943. Petroleum acids. VI. Naphthenic acids from California petroleum. J. Am. Chem. Soc. 65 (5), 770–777.

Pashley, R.M., Karaman, M.E., 2004. Applied Colloid and Surface Chemistry. John Wiley & Sons, Inc., Chichester, UK.

Peters, K.E., Walters, C.C., Moldowan, J.M., 2005. The Biomarker Guide 2nd Edition. Volume 2: Biomarkers and Isotopes in Petroleum Exploration and Earth History. Cambridge University Press, Cambridge, UK.

Petkova, N., Angelova, M., Petkov, P., 2009. Establishing the reasons and type of enhanced corrosion in the crude oil atmospheric distillation unit. Petroleum Coal 51 (4), 286–292.

Piehl, R.L., 1960. Correlation of corrosion in a crude distillation unit with the chemistry of the crudes. Corrosion 16, 6.

Piehl, R.L., 1988. Naphthenic corrosion in crude distillation units. Mater. Perform. 27 (1), 37–43.

Pillon, L.Z., 2008. Interfacial Properties of Petroleum Products. CRC Press, Taylor & Francis Group, Boca Raton, FL.

Poteau, S., Argillier, J.F., Langevin, D., Pincet, F., Perez, E., 2005. Influence of pH on stability and dynamic properties of asphaltenes and other amphiphilic molecules at the oil–water interface. Energy Fuels 19 (4), 1337–1341.

Qian, K., Robbins, W.K., Hughey, C.A., Cooper, H.J., Rodgers, R.P., Marshall, A.G., 2001. Resolution and identification of elemental composition for more than 3000 crude acids in heavy petroleum by negative-ion microelectrospray high-field Fourier transform ion cyclotron resonance mass spectrometry. Energy Fuels 15 (6), 1505–1511.

Qian, K., Edwards, K.E., Dechert, G.J., Jaffe, S.B., Green, L.A., Olmstead, W.N., 2008. Measurement of total acid number (TAN) and TAN boiling point distribution in petroleum products by electrospray ionization mass spectrometry. Anal. Chem. 2008 (80), 849–855.

Raia, J.C. < http://pubs.acs.org/action/doSearch?action=search&author=Hart%2C+H.+V.&qs SearchArea=author >.

Reinsel, M.A., Borkowski, J.J., Sears, J.T., 1994. Partition coefficients for acetic, propionic and butyric acids in a crude oil/water system. J. Chem. Eng. Data 39 (3), 513.

Rikka, P., 2007. Spectrometric Identification of Naphthenic Acids Isolated from Crude Oil (M.Sc. thesis). Department of Chemistry and Biochemistry Texas State University-San Marcos.

Robbins, W.K., 1998. Challenges in the characterization of naphthenic acids in petroleum. In: Proceedings of 215th National Meeting of American Chemical Society, pp. 137–140.

Rodgers, R.P., Hughey, C.A., Hendrickson, C.L., Marshall, A.G., 2002. Advanced characterization of petroleum crude and products by high field Fourier transform ion cyclotron resonance mass spectrometry. Div. Fuel Chem., Am Chem. Soc. 47 (2), 636–637.

Rogers, V., Liber, K., MacKinnon, M.D., 2002. Isolation and characterization of naphthenic acids from Athabasca oil sands tailings pond water. Chemosphere 48, 519–527.

Rogers, V.V., Wickstrom, M., Liber, K., MacKinnon, M.D., 2001. Acute and subchronic mammalian toxicity of naphthenic acids from oil sands tailings. Toxicol. Sci. 66 (2), 347–355.

Rostad, C.E., Hostettler, F.D., 2007. Profiling refined hydrocarbon fuels using polar components. Environ. Forensics 8, 129–137.

Saad, O.M., Gasmelseed, G.A., Hamid, A.H.M., 2014. Separation of naphthenic acid from Sudanese crude oil using local activated clay. J. Appl. Ind. Sci. 2 (1), 14–18.

Scattergood, G.L., Strong, R.C., 1987. Naphthenic acid corrosion, an update of control methods. Paper No. 197. In: Proceedings of the CORROSION'87. NACE International, Houston, TX.

Scott, A.C., Mackinnon, M.D., Fedorak, 2005. Naphthenic acids in Athabasca oil sands tailings waters are less biodegradable than commercial naphthenic acids. Environ. Sci. Technol. 39, 8388–8394.

Seifert, W.K., 1973. Steroid acids in petroleum—animal contribution to the origin of petroleum. Pure Appl. Chem. 34 (3–4), 633–640.

Seifert, W.K., Howells, W.G., 1969. Interfacially active acids in a California crude oil. Isolation of carboxylic acids and phenols. Anal. Chem. 41 (4), 554–562.

Shafizadeh, A., McAteer, G., Sigmon, J. 2010. High acid crudes. In: Crude Oil Quality Group—New Orleans Meeting, New Orleans, LA. Available from: <http://www.coqa-inc.org/20030130High%20Acid%20Crudes.pdf>.

Sheridan, M., 2006. California crude oil production and imports. Staff Paper No. CEC-600-2006-006. Fossil Fuels Office, Fuels and Transportation Division, California Energy Commission, Sacramento, CA.

Sjöblom, J., Aske, N., Auflem, I.H., Brandal, Ø., Havre, T.E., Sæther, Ø., et al., 2003. Our current understanding of water-in-crude oil emulsions. recent characterization techniques and high pressure performance. Adv. Colloid Interfacial Sci. 100–102, 399–473.

Somerville, A.C. <http://pubs.acs.org/action/doSearch?action=search&author=Raia%2C+J.+C.&qsSearchArea=author>.

Speight, J.G., 2014. The Chemistry and Technology of Petroleum, fifth ed. CRC Press, Taylor & Francis Group, Boca Raton, Florida.

Speight, J.G., 2014a. The Chemistry and Technology of Petroleum, fifth ed. CRC Press, Taylor & Francis Group, Boca Raton, FL.

Speight, J.G., 2014b. Oil and Gas Corrosion: From Surface Facilities to Refineries. Gulf Professional Publishing, Elsevier, Oxford, UK.

Speight, J.G., Arjoon, K.K., 2012. Bioremediation of Petroleum and Petroleum Products. Scrivener Publishing, Salem, MA.

Speight, J.G., Ozum, B., 2002. Petroleum Refining Processes. Marcel Dekker Inc., New York, NY.

Statoil, 2012. Grane Crude Oil Assay. Available from: <http://www.statoil.com/en/OurOperations/TradingProducts/CrudeOil/Crudeoilassays/Pages/Grane.aspx>.

Tai, X., Xian, S., 2011. Review and comprehensive analysis of composition and origin of high acidity crude oils. China Petrol. Process. Petrochem. Technol. 13 (1), 6–15.

Tebbal, S., Kane, R.D., 1996. Review of critical factors affecting crude corrosivity. In: Proceedings of the CORROSION'96. NACE International, Houston, TX.

Tebbal, S., Kane, R.D., Yamada, K, 1997. Assessment of the corrosivity of crude fractions from varying feedstocks. Paper No. 498. Proceedings. CORROSION/97. NACE International, Houston, Texas. (NB: the year should be «1997» in the citation) <http://pubs.acs.org/action/doSearch?action=search&author=Dzidic%2C+Ismet&qsSearchArea=author>.

Thorn, K.A., Aiken, G.R., 1998. Biodegradation of crude oil into nonvolatile organic acids in a contaminated Aquifer near Bemidji, Minnesota. Org. Geochem. 29 (4), 909–931.

Tomczyk, N.A., Winans, R.E., Shinn, J.H., Robinson, R.C., 2001. On the nature and origin of acidic species in petroleum. 1. Detailed acidic type distribution in a California crude oil. Energy Fuels 15 (6), 1498–1504.

Turnbull, A., Slavcheva, E., Shone, B., 1998. Factors controlling naphthenic acid corrosion. Corrosion 54 (11), 922–930.

US EPA., 2012. Naphthenic acids category analysis and hazard characterization. Submitted to the US EPA by The American Petroleum Institute Petroleum HPV Testing Group. Consortium Registration # 1100997. United States Environmental Protection Agency, Washington, DC, May 14.

Varadaraj, R., Brons, C., 2007. Molecular origins of heavy crude oil interfacial activity. Part 2: Fundamental interfacial properties of model naphthenic acids and naphthenic acids separated from heavy crude oils. Energy Fuels 21 (1), 199–204.

Vieth, A., Wilkes, H., 2006. Deciphering biodegradation effects on light hydrocarbons in crude oils using their stable carbon isotopic composition: a case study from the Gullfaks oil field, Offshore Norway. Geochim. Cosmochim. Acta 70 (3), 651–665.

Waples, D.W., 1985. Geochemistry in Petroleum Exploration. Reidel Publishing Company, Boston, MA.

Watson, J.S., Jones, D.M., Swannell, R.P.G., Van Duin, A.C.T., 2002. Formation of carboxylic acids during aerobic biodegradation of crude oil and evidence of microbial oxidation of hopanes. Org. Geochem. 33 (10), 1153–1169.

Wenger, L.M., Davis, C.L.,Isaksen, G.H., 2001. Multiple controls on petroleum biodegradation and impact on oil quality. Paper No. SPE 71450. In: Proceedings of the SPE Annual Technical Conference and Exhibition, New Orleans, September 30–October 3; also Paper No. 80168. SPE Reservoir Evaluation & Engineering. Society of Petroleum Engineers, Richardson, TX, 5(5): 375–383.

Wilkes, H., Kühner, S., Bolm, C., Fischer, T., Classen, A., Widdel, F., et al., 2003. Formation of n-alkane- and cycloalkane-derived organic acids during anaerobic growth of a denitrifying bacterium with crude oil. Org. Geochem. 34 (9), 1313–1323.

Wu, Q., 2010. High TAN Crude and Its Processing. CNOOC Refining & Marketing Group, Huizhou Refinery, Daya Bay, Huizhou City, Guangdong Province, China. <http://www.pecj.or.jp/japanese/overseas/conference/pdf/conference04-07.pdf>.

Yemashova, N.A., Murygina, V.P., Zhukov, D.V., Zakharyantz, A.A., Gladchenko, M.A., Appanna, V., et al., 2007. Biodeterioration of crude oil and oil derived products: a review. Rev. Environ. Sci. Biotechnol. 6 (4), 315–337.

Yeung, T.W., 2006. Evaluating opportunity crude processing. Petrol. Technol. Q. Q4, 93–97.

Zhao, B., Currie, R., Mian, H., 2012. Catalogue of analytical methods for naphthenic acids related to oil sands operations. OSRIN Report No. TR-21. Oil Sands Research and Information Network, University of Alberta, School of Energy and the Environment, Edmonton, AB, Canada.

CHAPTER 2

Amri, J., Gulbrandsen, E., Nogueira, R., 2011. Role of acetic acid in CO_2 top of the line corrosion of carbon steel. Paper No. 11329. In: Proceedings of the CORROSION/11. NACE International, Houston, TX.

ASTM D1386, 2012. Standard test method for acid number (empirical) of synthetic and natural waxes. Annual Book of Standards. ASTM International, West Conshohocken, PA.

ASTM D2896, 2012. Standard test method for base number of petroleum products by potentiometric perchloric acid titration. Annual Book of Standards. ASTM International, West Conshohocken, PA.

ASTM D3242, 2012. Standard test method for acidity in aviation turbine fuel. Annual Book of Standards. ASTM International, West Conshohocken, PA.

ASTM D3339, 2012. Standard test method for acid number of petroleum products by semi-micro color indicator titration. Annual Book of Standards. ASTM International, West Conshohocken, PA.

ASTM D4739, 2012. Standard test method for base number determination by potentiometric hydrochloric acid titration. Annual Book of Standards. ASTM International, West Conshohocken, PA.

ASTM D5770, 2012. Standard test method for semi-quantitative micro determination of acid number of lubricating oils during oxidation testing. Annual Book of Standards. ASTM International, West Conshohocken, PA.

ASTM D7253, 2012. Standard test method for polyurethane raw materials: determination of acidity as acid number for polyether polyols. Annual Book of Standards. ASTM International, West Conshohocken, PA.

ASTM D7389, 2012. Standard test method for acid number (empirical) of maleic anhydride (MAH) grafted waxes. Annual Book of Standards. ASTM International, West Conshohocken, PA.

ASTM D974, 2012. Standard test method for acid and base number by color-indicator titration. Annual Book of Standards. ASTM International, West Conshohocken, PA.

Ayello, F., Robbins, W., Richter, S., Nešić, S. 2011. Crude oil chemistry effects on inhibition of corrosion and phase wetting. Paper No. 11060. In: Proceedings of the Corrosion 2011. NACE International, Houston, TX.

Babaian-Kibala, E., Michael J., Nugent, M.J. 1999. Naphthenic acid corrosion literature survey. In: Proceedings of the CORROSION/99, NACE International, Houston, TX.

Backensto, E.B., Drew, R.E., Stapleford, C.C., 1956. High temperature hydrogen sulfide corrosion. Corrosion—NACE 12, 22.

Beavers, J.A., Thompson, N.G., 2006. External corrosion of oil and natural gas pipelines. ASM Handbook, Volume 13C, Corrosion: Environments and Industries. ASM International, Materials Park, OH.

Blanco, F., Hopkinson, B., 1983. Experience with naphthenic acid corrosion in refinery distillation process units. Paper No. 99. In: Proceedings of the CORROSION 83, NACE International, Houston, TX.

Bota, G., Nesic, S., 2013. Naphthenic acid challenges to iron sulfide scales generated *in-situ* from model oils on mild steel at high temperature. Paper No. 2512. In: Proceedings of the CORROSION/13. NACE International, Houston, TX.

Bradford, S.A., 1993. Corrosion Control. Van Nostrand Reinhold, New York, NY.

Bushman, J.B., 2002. Corrosion and Cathodic Protection Theory. Bushman & Associates Inc, Medina, OH.

Craig, H.L., 1995. Naphthenic acid corrosion in the refinery. Paper No. 333. In: Proceedings of the CORROSION/95. NACE International, Houston, TX.

Craig, H.L., 1996. Temperature and velocity effects in naphthenic acid corrosion. Paper No. 603. In: Proceedings of the CORROSION/96. NACE International, Houston, TX.

Derungs, W.A., 1956. Naphthenic acid corrosion—an old enemy of the petroleum industry. Corrosion 12 (2), 41.

Dettman, H.D., Li, N., Wickramasinghe, D., Luo, J. 2010. The influence of naphthenic acid and sulfur compound structure on global crude corrosivity under vacuum distillation conditions. In: Proceedings of the COQA/CCQTA Joint Meeting, New Orleans, LA.

Dougherty, J.A. 2004. A review of the effect of organic acids on CO_2 corrosion. Paper No. 04376. In: Proceedings of the CORROSION/04. NACE International, Houston, TX.

Fan, T.P., 1991. Characterization of naphthenic acids in petroleum by fast atom bombardment mass spectrometry. Energy Fuels 5 (3), 371–375.

Flego, C., Galasso, L., Montanari, L., Gennaro, M.E., 2013. Evolution of naphthenic acids during the corrosion process. Energy Fuels 27 (12). Available from: http://dx.doi.org/10.1021/ef401973z.

Fontana, M.G., 1986. Corrosion Engineering, third ed. McGraw-Hill, New York, NY.

Garsany, Y., Pletcher, D., Hedges, B. 2002. The role of acetate in CO_2 corrosion of carbon steel: has the chemistry been forgotten? Paper No. 02273. In: Proceedings of the CORROSION/02. NACE International, Houston, TX.

Garverick, L. (Ed.), 1994. Corrosion in the Petrochemical Industry. ASM International, Materials Park, OH.

Gary, J.G., Handwerk, G.E., Kaiser, M.J., 2007. Petroleum Refining: Technology and Economics, fifth ed. CRC Press, Taylor & Francis Group, Boca Raton, FL.

Gorbaty, M.L., Martella, D.J., Sartori, G., Savage, D.W., Ballinger, B.H., Blum, S.C., et al. 2001. Process for Neutralization of Petroleum Acids Using Overbased Detergents. United States Patent 6,054,042. April 25.

Gutzeit, J., 1977. Naphthenic acid corrosion in oil refineries. Mater. Perform. 33 (10), 24–35.

Hau, J.L., Mirabal, E.J. 1996. Experience with processing high sulfur naphthenic acid containing heavy crude oils. Paper No. LA96037. In: Proceedings of the 2nd NACE Latin American Region Corrosion Congress. NACE International, Houston, TX.

Heller, J.J., Merick, R.D., Marquand, E.B., 1963. Corrosion of refinery equipment by naphthenic acid. Mater. Prot. 2 (9), 44.

Hilton, N.P., Scattergood, G.L., 2010. Mitigate corrosion in your crude unit. Hydrocarb. Process. 92 (9), 75–79.

Hsu, C.S., Robinson, P.R. (Eds.), 2006. Practical Advances in Petroleum Processing, vols. 1,2. Springer Science, New York, NY.

Hucińska, J., 2006. Influence of sulfur on high temperature degradation of steel structures in the refinery industry. Adv. Mater. Sci. 6 (1), 16–25.

Hurlen, T., Gunvaldsen, S., 1984. Effects of carbon dioxide on reactions at iron electrodes in aqueous salt solutions. J. Electroanal. Chem. 180, 511–526.

Jayaraman, A., Singh, H., Lefebvre, Y., 1986. Naphthenic acid corrosion in petroleum refineries. A review. Revue Institut Français du Petrole 41, 265–274.

Jones, D.A., 1996. Principles and Prevention of Corrosion, second ed. Prentice Hall, Upper Saddle River, NJ.

Kane, R.D., Cayard, M.S., 1999. Understanding critical factors that influence refinery crude corrosiveness. Mater. Perform. 38 (7), 48.

Kane, R.D., Cayard, M.S., 2002. A comprehensive study on naphthenic acid corrosion. Paper No. 02555. In: Proceedings of the CORROSION/2002. NACE International, Houston, TX.

Kanukuntla, V., Qu, D.G., Nesic, S.R., Wolf, A., 2008. Experimental study of concurrent naphthenic acid and sulfidation corrosion. Paper No. 2764. In: Proceedings of the 17th International Congress, Las Vegas, NV. October 6–10. NACE International, Houston, TX.

Kittrell, N., 2006. Removing acid from crude oil. In: Proceedings of the Crude Oil Quality Group Meeting, New Orleans, LA. February.

Landolt, D., 2007. Corrosion and Surface Chemistry of Metals. CRC Press, Taylor & Francis Group, Boca Raton, FL.

Laredo, G.C., Lopez, C.R., Alvarez, R.E., Cano, J.L., 2004. Naphthenic acids, total acid number and sulfur content profile characterization in Isthmus and Maya crude oils. Fuel 83, 1689–1695.

Lewis K.R., Daane M.L., Schelling R., 1999. Processing corrosive crude oils. In: Proceedings of the CORROSION/99. NACE International, Houston, TX.

Linter, B., Burstein, G., 1999. Reactions of pipeline steels in carbon dioxide solutions. Corros. Sci. 41, 117–139.

Matos, M., Canhoto, C., Bento, M., Geraldo, M., 2010. Simultaneous evaluation of the dissociated and undissociated acid concentrations by square wave voltammetry using microelectrodes. J. Electroanal. Chem. 647, 144–149.

Meredith, W., Kelland, S.J., Jones, D.M., 2000. Influence of biodegradation on crude oil activity and carboxylic acid composition. Org. Geochem. 31 (11), 1059–1073.

Meriem-Benziane, M., Hamou Zahloul, H., 2013. Effect of corrosion on hydrocarbon pipelines. World Acad. Sci. Eng. Technol. 75, 285–287.

Messer, B. Tarleton, B., Beaton, M., Phillips, T., 2004. New theory for naphthenic acid corrosivity of Athabasca oil sands crudes. Paper No. 04634. In: Proceedings of the CORROSION 2004. pp. 1–11.

Peabody, A.W., 2001. Control of Pipeline Corrosion, second ed. NACE International, Houston, TX.

Petkova, N., Angelova, M., Petkov, P., 2009. Establishing the reasons and type of enhanced corrosion in the crude oil atmospheric distillation unit. Petrol. Coal 51 (4), 286–292.

Piehl, R.L., 1987. Naphthenic acid corrosion in crude oil distillation units. Paper No. 196. In: Proceedings of the CORROSION/87, Houston, TX.

Qu, D.R., Zheng, Y.G., Jing, H.M., Jiang, X., Ke, W., 2005. Erosion-Corrosion of Q235 and 5Cr1/2Mo steels in oil with naphthenic acid and/or sulfur compound at high temperature. Mater. Corros. 56 (8), 533–541.

Qu, D.R., Zheng, Y.G., Jing, H.M., Yao, Z.M., Ke, W., 2006. High temperature naphthenic acid corrosion and sulfidic corrosion of Q235 and 5Cr1/2Mo steels in synthetic refining media. Corros. Sci. 48, 1960–1985.

Qu, D.R., Zheng, Y.G., Jiang, X., Ke, W., 2007. Correlation between the corrosivity of naphthenic acids and their chemical structures. Anti-Corros. Methods Mater. 54 (4), 211–218.

Rebak, R.B., 2011. Sulfidic corrosion in refineries—a review. Corros. Rev. 29 (3–4), 123–134.

Remita, E., Tribollet, B., Sutter, E., Vivier, V., Ropital, F., Kittel, J., 2008. Hydrogen evolution in aqueous solutions containing dissolved CO_2: quantitative contribution of the buffering effect. Corros. Sci. 50, 1433–1440.

Scattergood, G.L., Strong, R.C., 1987. Naphthenic acid corrosion, an update of control methods. Paper No. 197. In: Proceedings of the CORROSION/87. NACE International, Houston, TX.

Shalaby, H.M., Al-Hashem, A., Lowther, M., Al-Besharah, J. (Eds.), 1996. Industrial Corrosion and Corrosion Control Technology. Kuwait Institute for Scientific Research, Safat, Kuwait.

Shreir, L.L., Jarman, R.A., Burstein, G.T. (Eds.), 1994. Corrosion, vols. 1,2. Butterworth–Heinemann, Oxford, UK.

Slavcheva, E., Shone, B., Turnbull, A., 1998. Factors controlling naphthenic acid corrosion. Paper No. 98579. In: Proceedings of the CORROSION/98. NACE International, Houston, TX.

Slavcheva, E., Shone, B., Turnbull, A., 1999. Review of naphthenic acid corrosion in oil refining. Br. Corros. J. 34 (2), 125–131.

Speight, J.G., 2009. Enhanced Recovery Methods for Heavy Oil and Tar Sands. Gulf Publishing Company, Houston, TX.

Speight, J.G., 2013a. The Chemistry and Technology of Coal, third ed. CRC Press, Taylor & Francis Group, Boca Raton, FL.

Speight, J.G., 2013b. Coal-Fired Power Generation Handbook. Scrivener Publishing, Salem, MA.

Speight, J.G., 2014a. The Chemistry and Technology of Petroleum, fifth ed. CRC Press, Taylor & Francis Group, Boca Raton, FL.

Speight, J.G., 2014b. Oil and Gas Corrosion: From Surface Facilities to Refineries. Gulf Professional Publishing, Elsevier, Oxford, UK.

Speight, J.G., Francisco, M.A., 1990. Studies in petroleum composition IV: changes in the nature of chemical constituents during crude oil distillation. Rev. de l'Institut Français du Pétrole 45, 733.

Speight, J.G., Ozum, B., 2002. Petroleum Refining Processes. Marcel Dekker Inc., New York, NY.

Tebbal, S., 1999. Critical review of naphthenic acid corrosion. Paper No. 380. In: Proceedings of the CORROSION/99. NACE International, Houston, TX.

Tebbal, S., Kane, R.D., 1996. Review of critical factors affecting crude corrosivity. Paper No. 607. In: Proceedings of the CORROSION/96. NACE International, Houston, TX.

Tebbal, S., Kane, R.D., 1998. Assessment of crude oil corrosivity. Paper No. 578. In: Proceedings of the CORROSION/98. NACE International, Houston, TX.

Tebbal, S., Kane, R.D., Yamada. K., 1997. Assessment of the corrosivity of crude fractions from varying feedstocks. Paper No. 498. In: Proceedings of the CORROSION/97. NACE International, Houston, TX.

Tran, T., Brown, B., Nesic, S., Tribollet, B., 2013. Investigation of the mechanism for acetic acid corrosion of mild steel. Paper No. 02487. In Proceedings of the CORROSION/13. NACE International, Houston, TX.

Uhlig, H.H., Revie, R.W., 1985. Corrosion and Corrosion Control: An Introduction to Corrosion Science and Engineering, third ed. John Wiley & Sons Inc., New York, NY.

Wang, C., Wang, Y., Chen, J., Sun, X., Liu, Z., Wan, Q., et al., 2011a. High temperature naphthenic acid corrosion of typical steels. Canadian J. Mech. Sci. Eng. 2 (2), 23–29.

Wang, Z., Li, B., Du, H., 2011b. Influence of naphthenates on the corrosion caused by naphthenic acid. Acta Petrolei Sin. (Petroleum Processing Section) 27 (3), 461–464.

Wranglen, G., 1985. An Introduction to Corrosion and Protection. Chapman & Hall, London, UK.

Wu, X.Q., Jing, H.M., Zheng, Y.G., Yao, Z.M., Ke, W., 2004a. Erosion-corrosion of various oil-refining materials in naphthenic acid. Wear 256, 133–144.

Wu, X.Q., Jing, H.M., Zheng, Y.G., Yao, Z.M., Ke, W., 2004b. Study on high-temperature naphthenic acid corrosion and erosion-corrosion of aluminized carbon steel. J. Mater. Sci. 39, 975–985.

Yépez, O., 2005. Influence of different sulfur compounds on corrosion due to naphthenic acid. Fuel 84 (1), 97–104.

Zetlmeisl, M.J., Harrel, J.B., Campbell, J., 2000. Naphthenic acid corrosion control. Hydrocarb. Eng.(3), 41–45.

CHAPTER 3

Acevedo, S., Escobar, G., Ranaudo, M.A., Khazen, J., Borges, B., Pereira, J.C., et al., 1999. Isolation and characterization of low and high molecular weight acidic compounds from cerro negro extra-heavy crude oil. Role of these acids in the interfacial properties of the crude oil emulsions. Energy Fuels 13, 333–335.

Babaian-Kibala, E., Nugent, M.J., 1999. Naphthenic acid corrosion: literature survey. Paper No. 378. In: Proceedings of the CORROSION'99. NACE International, Houston, TX.

Babaian-Kibala, E., Craig, H.L., Rusk, G.L., Blanchard, K.V., Rose, T.J., Uehlein, B.L., et al., 1993a. Naphthenic acid corrosion in refinery settings. Mater. Perform. 32 (4), 50–55.

Babaian-Kibala, E., Craig, H.L., Rusk, G.L., 1993b. In: Proceedings of CORROSION'93, New Orleans, LA. NACE International, Houston, TX, p. 631.

Braden, V.K., Malpiedi, M., Bowerbank, L., Gorman, J.P., 2007. Crude unit overhead corrosion control. Paper No. 98585. In: Proceedings of the CORROSION'07. NACE International, Houston, TX, 1998.

Craig, H.L., 1995. Naphthenic acid corrosion in the refinery. Paper No. 333. In: Proceedings of the CORROSION'95. NACE, International, Houston, TX.

Craig, H.L., 1996. Temperature and velocity effects in naphthenic acid corrosion. Paper No. 603. In: Proceedings of the CORROSION'96. NACE International, Houston, TX.

Derungs, W.A., 1956. Naphthenic acid corrosion—an old enemy of the petroleum industry. Corrosion 12 (2), 617–622.

Dzidic, I., Somerville, A.C., Raia, J.C., Hart, H.V., 1988. Determination of naphthenic acids in California crudes and refinery waste waters by fluoride ion chemical ionization mass spectrometry. Anal. Chem. 60, 1318–1323.

Ese, M.H., Kilpatrick, P.K., 2004. Stabilization of water-in-oil emulsions by naphthenic acids and their salts: model compounds, role of pH, and soap/acid ratio. J. Dispers. Sci. Technol. 25, 253–261.

Fan, T.P., 1991. Characterization of naphthenic acids in petroleum by fast atom bombardment mass spectrometry. Energy Fuels 5, 371–375.

Goldszal, A., Bourrel, M., Hurtevent, C., Volle, J.-L., 2002. Stability of water in acidic crude oil emulsions. In: Proceedings of the Third International Conference on Petroleum Phase Behavior and Fouling, New Orleans, LA.

Gorbaty, M.L., Martella, D.J., Sartori, G., Savage, D.W., Ballinger, B.H., Blum, S.C., et al., 2001. Process for Neutralization of Petroleum Acids Using Overbased Detergents. US Patent 6,054,042, April 25.

Gutzeit, J., 1976a. Naphthenic acid corrosion. Paper No. 156. In: Proceedings of the CORROSION'76. NACE International, Houston, TX.

Gutzeit, J., 1976b. Studies shed light on naphthenic acid corrosion. Oil Gas J. 74, 156.

Gutzeit, J., 1977. Naphthenic acid corrosion in oil refineries. Mater. Perform. 33 (10), 24.

Heller, J.J., Merick, R.D., Marquand, E.B., 1963. Corrosion of refinery equipment by naphthenic acid. NACE Publication 8B163, Materials Protection, 2(9), 44. NACE International, Houston, TX.

Hopkinson, B.R., Penuela, L.E., 1997. Naphthenic acid corrosion by Venezuelan crudes. Paper No. 502. In: Proceedings of the CORROSION'97. NACE International, Houston, TX.

Hsu, C.S., Dechert, G.J., Robbins, W.K., Fukuda, E.K., 2000. Naphthenic acids in crude oils characterized by mass spectrometry. Energy Fuels 14, 217–223.

Jayaraman, A., Singh, H., Lefebvre, Y., 1986. Naphthenic acid corrosion in petroleum refineries. A review. Rev. I. Fr. Petrol. 41, 265–274.

Johnson, D., McAteer, G., Zuk, H., 2003a. The safe processing of high naphthenic acid content crude oils refinery experience and mitigations studies. Paper No. 03645. In: CORROSION'03. NACE International, Houston, TX.

Johnson, D., McAteer, G.R., Zuk, H., 2003b. Mitigating corrosion from naphthenic acid streams. Petrol. Technol. Q. Q1, 79–86.

Kane, R.D., Cayard, M.S., 1999. Understanding critical factors that influence refinery crude corrosiveness. Mater. Perform. 38 (7), 48.

Kane, R.D., Cayard, M.S., 2002. A comprehensive study on naphthenic acid corrosion. Paper No. 02555. In: CORROSION'02. NACE International, Houston, TX.

Kapusta, S.D., Rinus Daane, F., Place, M.C., 2003. The impact of oil field chemicals on refinery corrosion problems. Paper No. 03649. In: Proceedings of the CORROSION'07. NACE International, Houston, TX.

Kapusta, S.D., Ooms, A., Smith, A., Van den Berg, F., Fort, W., 2004. Safe processing of acid crudes. Paper No. 04637. In: Proceedings of the CORROSION'04. NACE International, Houston, TX.

Kittrell, N., 2006. Removing acid from crude oil. Crude Oil Quality Group. In: New Orleans Meeting, New Orleans, LA, February.

Laredo, G.C., Lopez, C.R., Alvarez, R.E., Cano, J.L., 2004. Naphthenic acids, total acid number and sulfur content profile characterization in isthmus and maya crude oils. Fuel 83, 1689–1695.

Lewis, K.R., Daane, M.L., Schelling, R., 1999. Processing corrosive crude oils. Paper No. 377. In: Proceedings of the CORROSION'99. NACE International, Houston, TX.

Lordo, S., Garcia, J.M., Garcia-Swofford, S., 2008. Desalter acidification additives and their potential impacts on crude units. Paper No. 08556. In: Proceedings of the CORROSION'08. NACE International, Houston, TX.

Messer, B. Tarleton, B., Beaton, M., Phillips, T., 2004. New theory for naphthenic acid corrosivity of athabasca oil sands crudes. Paper No. 04634. In: Proceedings of the CORROSION'04. NACE International, Houston, TX.

Morrison B.L., DeAngelis, D., Bonnette, L., Wood, S., 1992. The determination of naphthenic acids in crude oil. In: Proceedings of the PITTCON'92, New Orleans, LA.

Nugent, M.J., Dobis, J.D., 1998. Experience with naphthenic acid corrosion in low tan crudes. Paper No. 577. In: Proceedings of the CORROSION'98. NACE International, Houston, TX.

Pathak, A.K., Kumar, T., 1995. Study of indigenous crude oil emulsions and their stability. In: Proceedings of the PETROTECH'95. Technology trends in the Oil Industry. Ministry of Petroleum and Natural Gas, New Delhi, India.

Piehl, R.L., 1960. Correlation of corrosion in a crude distillation unit with chemistry of the crudes. Corrosion 16, 6.

Piehl, R.L., 1988. Naphthenic acid corrosion in crude distillation units. Mater. Perform. 44 (1), 37–43.

Qu, D.R., Zheng, Y.G., Jing, H.M., Jiang, X., Ke, W., 2005. Erosion–Corrosion of Q235 and 5Cr1/2Mo steels in oil with naphthenic acid and/or sulfur compound at high temperature. Mater. Corrosion 56 (8), 533–541.

Qu, D.R., Zheng, Y.G., Jing, H.M., Yao, Z.M., Ke, W., 2006. High temperature naphthenic acid corrosion and sulfidic corrosion of Q235 and 5Cr1/2Mo steels in synthetic refining media. Corrosion Sci. 48, 1960–1985.

Qu, D.R., Zheng, Y.G., Jiang, X., Ke, W., 2007. Correlation between the corrosivity of naphthenic acids and their chemical structures. Anti-Corrosion Methods Mater. 54 (4), 211–218.

Rue, J.R., Naeger, D.P., 2007. Advances in crude unit corrosion control. Paper No. 199. In: Proceedings of the CORROSION'07. NACE International, Houston, TX.

Seifert, W.K., Teeter, R.M., 1970. Identification of polycyclic aromatic and heterocyclic crude oil carboxylic acids. Anal. Chem. 42, 750–758.

Shalaby, H.M., 2005. Refining of Kuwait's heavy crude oil: materials challenges. In: Proceedings of the Workshop on Corrosion and Protection of Metals. Arab School for Science and Technology, Kuwait, December 3–7.

Shargay, C., Moore, K., Colwell, R., 2007. Survey of materials in hydrotreater units processing high TAN feeds. Paper No. 07573. In: Proceedings of the CORROSION'07. NACE International, Houston, TX.

Slater, J.E., Berry, W.R., Paris, B., Boyd, W.K., 1974. High temperature crude oil corrosivity studies. API Publication No. 943, American Petroleum Institute, Washington, DC, September.

Slavcheva, E., Shone, B., Turnbull, A., 1998. Factors controlling naphthenic acid corrosion. Paper No. 98579. In: Proceedings of the CORROSION'98. NACE International, Houston, TX.

Slavcheva, E., Shone, B., Turnbull, A., 1999. Review of naphthenic acid corrosion in oil refining. Br. Corrosion J. 34 (2), 125–131.

Speight, J.G., 2001. Handbook of Petroleum Analysis. John Wiley & Sons, Inc., New York, NY.

Speight, J.G., 2002. Handbook of Petroleum Product Analysis. John Wiley & Sons, Inc., Hoboken, NJ.

Speight, J.G., 2014a. The Chemistry and Technology of Petroleum, fifth ed. CRC Press, Taylor & Francis Group, Boca Raton, FL.

Speight, J.G., 2014b. Oil and gas corrosion: from surface facilities to refineries. Gulf Professional Publishing. Elsevier, Oxford, UK.

Speight, J.G., Francisco, M.A., 1990. Studies in petroleum composition IV: changes in the nature of chemical constituents during crude oil distillation. Rev. I. Fr. Pétrol. 45, 733.

Tebbal, S., 1999. Critical review on naphthenic acid corrosion. Paper No. 380. In: Proceedings of the CORROSION'99. NACE International, Houston, TX.

Tebbal, S., Kane, R.D., Yamada, K., 1997. Assessment of the corrosivity of crude fractions from varying feedstocks. Paper No. 498. In: Proceedings of the CORROSION'97. NACE International, Houston, TX.

Tran, T., Brown, B., Nesic, S., Tribollet, B., 2013. Investigation of the mechanism for acetic acid corrosion of mild steel. Paper No. 2487. In: Proceedings of the CORROSION'13. NACE International, Houston, TX.

Turnbull, A., Slavcheva, E., Shone, B., 1994. Factors controlling naphthenic acid corrosion. In: CORROSION'94. NACE International, Houston, TX.

Turnbull, A., Slavcheva, E., Shone, B., 1998. Factors controlling naphthenic acid corrosion. Corrosion 54 (11), 922–930.

UOP Method 565-92, 1992. Acid Number and Naphthenic Acids by Potentiometric Titration. UOP, Des Plaines, IL.

UOP Method 587-92, 1992. Acid Number and Naphthenic Acids by Colorimetric Titration. UOP, Des Plaines, IL.

Wang, C., Wang, Y., Chen, J., Sun, X., Liu, Z., Wan, Q., et al., 2011. High temperature naphthenic acid corrosion of typical steels. Can. J. Mech. Sci. Eng. 2 (2), 23–29.

Yépez, O., 2005. Influence of different sulfur compounds on corrosion due to naphthenic acid. Fuel 84 (1), 97–104.

Zetlmeisl, M.J., 1996. Naphthenic acid corrosion and its control. Paper No. 218. In: Proceedings of the CORROSION'96. NACE International, Houston, TX.

CHAPTER 4

Babaian-Kibala, E., Nugent, M.J., 1999. Naphthenic acid corrosion: literature survey. Paper No. 378. In: Proceedings of the CORROSION/99. NACE International, Houston, TX.

Babaian-Kibala, E., Craig, H.L., Rusk, G.L., Quinter, R.C., Summers, M.A., 1993. Naphthenic acid corrosion in refinery settings. Mater. Perform. 32 (4), 50–55.

Blanco, E.F., Hopkinson, B., 1983. Experience with naphthenic acid corrosion in refinery distillation process units. Paper No. 99. In: Proceedings of the CORROSION/83. NACE International, Houston, TX.

Craig, H.L., 1995. Naphthenic acid corrosion in the refinery. Paper No. 333. In: Proceedings of the CORROSION/95. NACE, International, Houston, TX.

Craig, H.L., 1996. Temperature and velocity effects in naphthenic acid corrosion. Paper No. 603. In: Proceedings of the CORROSION/96. NACE International, Houston, TX.

Dean, F.W.H., Powell, S.J., 2006. Hydrogen flux and high temperature acid corrosion. Paper No. 06436. In: Proceedings of the CORROSION/06. NACE International, Houston, TX.

Gary, J.G., Handwerk, G.E., Kaiser, M.J., 2007. Petroleum Refining: Technology and Economics, fifth ed. CRC Press, Taylor & Francis Group, Boca Raton, FL.

Groysman, A., Brodsky, N., Pener, J., Goldis, A., Savchenko, N., 2005. Study of corrosiveness of acidic crude oil and fractions. Paper No. 05686. In: Proceedings of the CORROSION 2005. NACE International, Houston, TX.

Groysman, A., Brodsky, N., Pener, J., Shmulevich, D., 2007. Low temperature naphthenic acid corrosion study. Paper No. 07569. In: Proceedings of the CORROSION/07. NACE International, Houston, TX.

Hau, J.L., Mirabal, E.J., 1996. Experience with processing high sulfur naphthenic acid containing heavy crude oils. Paper No. LA96037. In: Proceedings of the Second NACE Latin American Region Corrosion Congress. NACE International, Houston, TX.

Hopkinson, B.E., Penuela, L.E., 1997. Naphthenic acid corrosion by Venezuelan crudes. Paper No. 502. In: Proceedings of the CORROSION/97. NACE International, Houston, TX.

Hsu, C.S., Robinson, P.R. (Eds.), 2006. Practical Advances in Petroleum Processing Volume 1 and Volume 2. Springer Science, New York, NY.

Johnson, D., McAteer, G.R., Zuk, H., 2003. Mitigating corrosion from naphthenic acid streams. Petrol. Technol. Quart. Q1, 79–86.

Kapusta, S.D., Ooms, A., van den Berg, F., Smith, A., Fort, W.C., 2004. Safe processing of acid crudes. Paper No. 04637. In: Proceedings of the CORROSION/04. NACE International, Houston, TX.

Mottram, R.A., Hathaway, J.T., 1971. Some experience in the corrosion of a crude oil distillation unit operating with low sulfur North African crudes. Paper No. 39. In: Proceedings of the CORROSION/71. NACE International, Houston, TX.

Nugent, M.J., Dobis, J.D., 1998. Experience with naphthenic acid corrosion in low TAN Crudes. Paper No. 577. In: Proceedings of the CORROSION/98. NACE International, Houston, TX.

O'Kane, J.M., Rudd, T.F., Harrison, J.H., Powell, S.W., Dean, F.W.H., 2010a. Correlation of hydrogen flux and corrosion rate measurements carried out during a severe episode of corrosion-erosion attributable to naphthenic acid. Paper No. 10178. In: Proceedings of the CORROSION/10. NACE International, Houston, TX.

O'Kane, J.M., Rudd, T.F., Cooke, D., Dean, F.W.H., Powell, S.W., 2010b. Detection and monitoring of naphthenic acid corrosion in a visbreaker unit using hydrogen flux measurements. Paper No. 10351. In: Proceedings of the CORROSION 2010, NACE International, Houston, TX.

Petkova, N., Angelova, M., Petkov, P., 2009. Establishing the reasons and type of enhanced corrosion in the crude oil atmospheric distillation unit. Petrol. Coal 51 (4), 286−292.

Piehl, R.L., 1988. Naphthenic acid corrosion in crude distillation units. Mater. Perform. 27 (1), 37−43.

Radovanović, L., Speight, J.G., 2011. Visbreaking: a technology of the future. In: Proceedings of the First International Conference—Process Technology and Environmental Protection (PTEP 2011). University of Novi Sad, Technical Faculty "Mihajlo Pupin," Zrenjanin, Republic of Serbia. December 7, 2011. pp. 335−338.

Rijkaart, M., Vanacore, M., Russell, C., 2009. Visbreaker optimization: a step change. Petrol. Technol. Quart. Q2, 71−78.

Rudd, T.F., O'Kane, J.M., Harrison, J.H., Dean, F.W.H., Powell, S.J., 2010. Correlation of hydrogen flux and corrosion rate measurements carried out during a severe episode of corrosion-erosion attributable to naphthenic acid. Paper No. 10178. In: Proceedings of the CORROSION/10. NACE International, Houston, TX.

Speight, J.G., 2014a. The Chemistry and Technology of Petroleum, fifth ed. CRC Press, Taylor & Francis Group, Boca Raton, FL.

Speight, J.G., 2014b. Oil and Gas Corrosion: From Surface Facilities to Refineries. Gulf Professional Publishing, Elsevier, Oxford, UK.

Speight, J.G., Francisco, M.A., 1990. Studies in petroleum composition IV: changes in the nature of chemical constituents during crude oil distillation. Rev de l'Institut Français du Pétrole 45, 733.

Speight, J.G., Ozum, B., 2002. Petroleum Refining Processes. Marcel Dekker Inc., New York, NY.

Stark, J.L., Nguyen, J., Kremer, L.N., 2002. Crude stability as related to desalter upsets. In: Proceedings of the International Conference on Refinery Processing, AIChE Spring National Meeting, New Orleans, LA. March.

Tebbal, S., 1999. Critical review of naphthenic acid corrosion. Paper No. 380. In: Proceedings of the CORROSION/99. NACE International, Houston, TX.

Turini, K., Turner, J., Chu, A., Vaidyanathan, S., 2011. Processing heavy crudes in existing refineries. In: Proceedings of the AIChE Spring Meeting, Chicago, IL. American Institute of Chemical Engineers, New York, NY. <http://www.aiche-fpd.org/listing/112.pdf>.

Wu, X.Q., Jing, H.M., Zheng, Y.G., Yao, Z.M., Ke, W., 2004a. Resistance of Mo-bearing stainless steels and Mo-bearing stainless-steel coating to naphthenic acid corrosion and erosion-corrosion. Corros. Sci. 46, 1013−1032.

Wu, X.Q., Jing, H.M., Zheng, Y.G., Yao, Z.M., Ke, W., 2004b. Erosion-corrosion of various oil-refining materials in naphthenic acid. Wear 256, 133−144.

Zetlmeisl, M.J., 1996. Naphthenic acid corrosion and its control. Paper No. 218. In: Proceedings of the CORROSION/96. NACE International, Houston, TX.

CHAPTER 5

Alvisi, P., Lins, V., 2011. An overview of naphthenic acid corrosion in a vacuum distillation plant. Eng. Failure Anal. 18 (5), 1403–1406.

Biryukova, V., Fedorak, P., Quideau, S., 2007. Biodegradation of naphthenic acids by rhizosphere microorganisms. Chemosphere 67 (10), 2058–2064.

Bosmann, A., Datsevich, L., Jess, A., Lauter, A., Schmitz, C., Wasserscheid, P., 2001. Deep desulfurization of diesel fuel by extraction with ionic liquids. Chem. Commun., 2494–2495.

Bradford, S.A., 1993. Corrosion Control. Van Nostrand Reinhold, New York, NY.

Burlov, V.V., Altsybeeva, A.I., Kuzinova, T.M., 2013. A new approach to resolve problems in the corrosion protection of metals. Int. J. Corros. Scale Inhib. 2 (2), 92–101.

Clemente, J.S., Fedorak, P.M., 2005. A review of the occurrence, analyses, toxicity, and biodegradation of naphthenic acids. Chemosphere 60, 585–600.

Cross, C., 2013. High-acid crude processing enabled by unique use of computational fluid dynamics. Petrol. Technol. Quart. Q4, 39–49.

Danzik, M., 1987. Process for removing naphthenic acids from petroleum distillates. US Patent No. 4,634,519, January 6.

Darensbourg, D.J., Joo, F., Kannisto, M., Katho, A., Reibenspies, J.H., Daigle, D.J., 1994. The catalytic hydrogenation of aldehydes in an aqueous two-phase solvent system using a 1,3,5-Triaza-7-phospha-adamantane complex of ruthenium. Inorg. Chem. 33, 200–208.

Ding, L., Rahimi, P., Hawkins, R., Bhatt, S., Shi, Y., 2009. Naphthenic acid removal from heavy oils on alkaline earth-metal oxides and ZnO catalysts. Appl. Catal. A: General 371 (1–2), 121–130.

Duan, Y., Yu, F., Cui, Z., 2012. Naphthenic acid corrosion control strategies in refineries: a review of recent patents. Recent Patents Corros. Sci. 2 (2), 142–147.

Edmonson, J.G., 1987. Method of inhibiting propionic acid corrosion in distillation units. US Patent 4,647,366, March 3.

Ese, M.H., Peter, K., Kilpatrick, P.K., 2004. Stabilization of water-in-oil emulsions by naphthenic acids and their salts: model compounds, role of pH, and soap: acid ratio. J. Dispers. Sci. Technol. 25 (3), 253–261.

Gaikar, V.G., Maiti, D., 1996. Adsorptive recovery of naphthenic acids using ion-exchange resins. React. Funct. Polym. 31, 155–164.

Garverick, L., 1994. Corrosion in the Petrochemicals Industry. Materials Park, OH. ASM International.

Gary, J.G., Handwerk, G.E., Kaiser, M.J., 2007. Petroleum Refining: Technology and Economics, fifth ed. CRC Press, Taylor & Francis Group, Boca Raton, FL.

Gorbaty, M.L., Martella, D.J., Sartori, G., Savage, D.W., Ballinger, B.H., Blum, S.C., et al., 2000. Process for neutralization of petroleum acids using over-based detergents. US Patent No. 6,054,042, April 25.

Gordon, C.M., 2001. New developments in catalysis using ionic liquids. Appl. Catal. A 222 (1–2), 101–117.

Greaney, M.A., 2003. Method for reducing the naphthenic acid content of crude oil and fractions. US Patent No. 6,531,055, March 11.

Groysman, A., 1995. Corrosion monitoring in the oil refinery. Paper No. 07. In: Proceedings of the International Conference on Corrosion in Natural and Industrial Environments: Problems and Solutions, Italy, May 23–25.

Groysman, A., 1996. Corrosion monitoring in the oil refinery. In: Proceedings of the 13th International Corrosion Congress, Melbourne, Australia, November 25–29.

Groysman, A., 1997. Corrosion monitoring and control in refinery process unit. Paper No. 512. In: Proceedings of the CORROSION'97, New Orleans, LA.

Gürses, A., Dograr, Ç., Yalçin, M., Açikyildiz, M., Bayrak, R., Karaca, S., 2006. J. Hazard Mater. B131, 217–228.

Hansmeier, A.R., Meindersma, G.W., De Haan, A.B., 2011. Desulfurization and denitrogenation of gasoline and diesel fuels by means of ionic liquids. Green Chem. 13, 1907–1913.

Hau, J.X., Yépez, O.J., Torres, L.H., Vera, J.R., 2003. Measuring naphthenic acid corrosion potential with the fe powder. Test. Rev, Metal Madrid, Vol. Extr., 116–123.

Hsu, C.S., Robinson, P.R. (Eds.), 2006. Practical Advances in Petroleum Processing: Volume 1 and Volume 2. Springer Science, New York, NY.

Huang, M., Zhao, S., Li, P., Huisingh, D., 2006. Removal of naphthenic acid by microwaves. J. Clean. Prod. 14 (8), 736–739.

Huang, Y., Zhu, J., Wu, B., Wang, Y., 2009. Catalytic esterification and deacidification of crude oil with high naphthenic acid content. CTA Petrol. Sin. (Petroleum Processing Section) 25 (5), 731–735.

Johnson, D., McAteer, G.R., Zuk, H., 2003. Mitigating corrosion from naphthenic acid streams. Petrol. Technol. Quart. Q1, 79–86.

Jones, D.A., 1996. Principles and Prevention of Corrosion, second ed. Prentice Hall, Upper Saddle River, NJ.

Kamarudin, H., Abdul Mutalib, M.I., Man, Z., Azmi, M., 2012a. Extraction of naphthenic acids from liquid hydrocarbon using imidazolium ionic liquids. In: Proceedings of the 2012 International Conference on Environment Science and Engineering (IPCBEE), vol. 32, pp. 17–23.

Kamarudin, H., Abdul Mutalib, M.I., Man, Z., 2012b. Extraction of carboxylic acids from hydrocarbon mixture using imidazolium ionic liquids. Int. J. Biosci. Biochem. Bioinform. 2 (4), 243–247.

Kane, R.D., Cayard, M.S., 2002. A comprehensive study on naphthenic acid corrosion. Paper No. 02555. In: Proceedings of the CORROSION'02. NACE International, Houston, TX.

Knag, M., 2005. Fundamental behavior of model corrosion inhibitors. J. Dispers. Sci. Technol. 27, 587–597.

Kume, Y., Qiao, K., Daisuke, T., 2008. Selective hydrogenation of cinnamaldehyde catalyzed by palladium nanoparticles immobilized on ionic liquids modified-silica gel. Catal. Commun. 9 (3), 369–375.

Kędra-Krolik, K., Mutelet, F., Moïse, J.C., Jaubert, J.N., 2011. Deep fuels desulfurization and denitrogenation using 1-butyl-3-methylimidazolium trifluoromethanesulfonate. Energy Fuels 25, 1559–1565.

Liotta, R., 1981. Oxygen-alkylation of carbonous material and products thereof. US Patent 4,300,995, November 17.

Marták, J., Schlosser, S., 2007. Extraction of lactic acid by phosphonium ionic liquids. Sep. Purif. Technol. 57, 483–494.

Matsumoto, M., Mochiduki, K., Fukunishi, K., Kondo, K., 2004. Extraction of organic acids using imidazolium-based ionic liquids and their toxicity to *Lactobacillus rhamnosus*. Sep. Purif. Technol. 40, 97–101.

Mottram, R.A., Hathaway, J.T., 1971. Some experience in the corrosion of a crude oil distillation unit operating with low sulfur north African crudes. Paper No. 39. In: Proceedings of the CORROSION'71. NACE International, Houston, TX.

Nie, Y., Li, C., Sun, A., Meng, H., Wang, Z., 2006. Extractive desulfurization of gasoline using imidazolium-based phosphoric ionic liquids. Energy Fuels 20, 2083–2087.

Norshahidatul Akmar, M.S., Jafariah, J., Wan Azelee, W.A.B., 2012. A chemical technique for total acid number reduction in crude oil. In: Proceedings of the UMT (Universiti Malaysia Terengganu) 11th International Annual Symposium on Sustainability Science and Management, Terengganu, Malaysia, July 9–11.

Norshahidatul Akmar, M.S., Wan Azelee, W.A.B., Jafariah, J., Shukri, N.M., 2013. Treatment of acidic petroleum crude oil utilizing catalytic neutralization technique of magnesium oxide catalyst. Mod. Chem. Appl. 1 (2), 1057–1062.

Ozvatan, S., Yurum, Y., 2002. Catalytic decarboxylation of Elbistan lignite. Energy Sources 24 (6), 581–589.

Pereira Silva, J., Ferreira de Senna, L., Baptista do Lago, D.C., Paulo Ferreira da Silva Jr., P., Gonçalves Dias, E., Gaya de Figueiredo, M.A., et al., 2007. Characterization of commercial ceramic adsorbents and its application on naphthenic acids removal of petroleum distillates. Mater. Res. 10 (2), 219–225.

Petersen, P.R., Robbins, F.P., Winston, W.G., 1993. Naphthenic acid corrosion inhibitors. US Patent 5,182,013, January 26.

Petkova, N., Angelova, M., Petkov, P., 2009. Establishing the reasons and type of enhanced corrosion in the crude oil atmospheric distillation unit. Petrol. Coal 51 (4), 286–292.

Piehl, R.L., 1960. Correlation of corrosion in a crude distillation unit with chemistry of the crudes. Corrosion 16, 6.

Rikka, P., 2007. Spectrometric Identification of Naphthenic Acids Isolated from Crude Oil (M.Sc. thesis). Department of Chemistry and Biochemistry Texas State University-San Marcos.

Saad, O.M., Gasmelseed, G.A., Hamid, A.H.M., 2014. Separation of naphthenic acid from Sudanese crude oil using local activated clays. J. Appl. Ind. Sci. 2 (1), 14–18.

Samuelson, G.J., 1954. Hydrogen-chloride evolution from crude oils as a function of salt concentration. In: Proceedings of the API, American Petroleum Institute, Washington, DC, vol. 34(III), pp. 50–54.

Sartori, G., Savage, D.W., Gorbaty, M.L., Ballinger, B.H., 1997. Process for neutralization of petroleum acids using alkali metal trialkylsilanolates. US Patent No. 5,643,439, July 1.

Sartori, G., Savage, D.W., Olmstead, W.N., Robbins, W.K., Dalrymple, D.C., Ballinger, B.H. 2001. Process for treatment of petroleum acids with ammonia. US Patent No. 6,258,258, July 10.

Sastri, V., 1998. Corrosion Inhibitors: Principles and Applications. John Wiley & Sons, Inc., Hoboken, NJ.

Scattergood, G.L., Strong, R.C., 1987. Naphthenic acid corrosion, an update of control methods. Paper No. 197. In: Proceedings of the CORROSION'87. NACE International, Houston, TX.

Shi, L., Wang, G., Shen, B., 2010. The removal of naphthenic acids from Beijiang crude oil with a sodium hydroxide solution of ethanol. Petrol. Sci. Technol. 28 (13), 1373–1380.

Shi, L.J., Shen, B.X., Wang, G.Q., 2008. Removal of naphthenic acids from Beijiang crude oil by forming ionic liquids. Energy Fuels 22, 4177–4181.

Simon, S., Nordgård, E., Bruheim, P., Sjöblom, J., 2008. Determination of C80 tetra-acid content in calcium naphthenate deposits. J. Chromatogr. A 1200 (2), 136–143.

Slavcheva, E., Shone, B., Turnbull, A., 1998. Factors controlling naphthenic acid corrosion. Paper No. 98579. In: Proceedings of the CORROSION'98. NACE International, Houston, TX.

Slavcheva, E., Shone, B., Turnbull, A., 1999. Review of naphthenic acid corrosion in oil refining. Br. Corros. J. 34 (2), 125–131.

Speight, J.G., 2009. Enhanced Recovery Methods for Heavy Oil and Tar Sands. Gulf Publishing Company, Houston, TX.

Speight, J.G., 2014a. The Chemistry and Technology of Petroleum, fifth ed. CRC Press, Taylor & Francis Group, Boca Raton, FL.

Speight, J.G., 2014b. Oil and Gas Corrosion Prevention: From Surface Facilities to Refineries. Gulf Professional Publishing, Elsevier, Oxford, UK.

Speight, J.G., Francisco, M.A., 1990. Studies in petroleum composition IV: changes in the nature of chemical constituents during crude oil distillation. Rev. I. Fr. Pétrol. 45, 733.

Speight, J.G., Ozum, B., 2002. Petroleum Refining Processes. Marcel Dekker Inc., New York, NY.

Varadaraj, R., Pugel, T.M., Savage, D.W., 1998. Removal of naphthenic acids in crude oils and distillates. US Patent 6,096,196, August 1.

Varadaraj, R., Pugel, T.M., Savage, D.W., 2000. Removal of naphthenic acids in crude oils and distillates. US Patent 6,096,196, August 1.

Verachtert, T.A., 1980. Trace acid removal in the pretreatment of petroleum distillate. US Patent 4,199,440, April 22.

Vetters, E., Clarida, D., 2013. Maintaining reliability when processing opportunity crudes. Petrol. Technol. Quart. Q4, 59–67.

Wang, Y., Chu, Z., Qiu, B., Liu, C., Zhang, Y., 2006. Removal of naphthenic acids from vacuum fraction oil with an ammonia solution of ethylene glycol. Fuel 85, 2489–2493.

Wang, Y., Sun, X., Liu, Y., Liu, C., 2007. Removal of naphthenic acids from a diesel fuel by esterification. Energy Fuels 21, 941–943.

Wang, Y., Liu, Y., Liu, C., 2008. Kinetics of the esterification of low-concentration naphthenic acids and methanol in oils with or without SnO as a catalyst. Energy Fuels 22 (4), 2203–2206.

White, R.A., Ehmke, E.F., 1991. Materials Selection for Refineries and Associated Facilities. NACE International, Houston, TX.

Yang, B., Xu, C.M., Zhao, S.Q., Hsu, C.S., Chung, K.H., Shi, Q., 2013. Thermal transformation of acid compounds in high TAN crude oil. Sci. China Chem. 56 (7), 848–855.

Zhang, A., Ma, Q., Tang, Y., 2004. Catalytic decarboxylation for naphthenic acid removal from crude oil—a theoretical and experimental study. Div. Petrol. Chem. Am. Chem. Soc. 49 (2), 218.

Zhang, A., Ma, Q., Wang, K., Tang, Y., Goddard, W.A., 2005. Improved processes to remove naphthenic acids. Final Technical Report. DOE Award number: DE-FC26-02NT15383. California Institute of Technology, Pasadena, CA, December 9.

Zhang, A., Ma, Q., Wang, K., Liu, X., Shuler, P., Tang, Y., 2006. Naphthenic acid removal from crude oil through catalytic decarboxylation on magnesium oxide. Appl. Catal. A: General 303 (1), 103–109.

Zou, L., Han, B., Yan, H., Kaperski, K.L., Xu, Y., Hepler, L.G., 1997. Enthalpy of adsorption and isotherms for adsorption of naphthenic acid onto clays. J. Colloid Interface Sci. 190, 472–475.

Printed and bound by CPI Group (UK) Ltd, Croydon, CR0 4YY

03/10/2024

01040423-0001